REVERSIBILITY AND STOCHASTIC NETWORKS

This classic in stochastic network modelling broke new ground when it was published in 1979, and it remains a superb introduction to reversibility and its applications.

The book concerns behaviour in equilibrium of vector stochastic processes or stochastic networks. When a stochastic network is reversible its analysis is greatly simplified, and the first chapter is devoted to a discussion of the concept of reversibility. The rest of the book focuses on the various applications of reversibility and the extent to which the assumption of reversibility can be relaxed without destroying the associated tractability. Now back in print for a new generation, this book makes enjoyable reading for anyone interested in stochastic processes thanks to the author's clear and easy-to-read style. Elementary probability is the only prerequisite and exercises are interspersed throughout.

'the exposition is clear and precise without being pedantic ... an example of the art of the true applied mathematician, who can penetrate the essence of a real problem by applying the right mathematical tool in just the right place. This book, then, will be essential (and enjoyable) reading for any operational researcher.'
J. F. C. Kingman
European Journal of Operational Research

'*Reversibility and Stochastic Networks* is simply a timeless classic. Students in operations research, electrical engineering, management science, mathematics, etc. can benefit tremendously by reading the technically deep and elegantly presented material in this book.'
Nick Bambos, Stanford University

'Its wealth of ideas is so rich and (as in much of Kelly's work) it starts with very elementary ideas which in his hands are built up until suddenly you have something that is very valuable.... It is a book that should be on the shelf of anyone working on stochastic networks.'
Onno Boxma, EURANDOM and Eindhoven University of Technology

REVERSIBILITY AND STOCHASTIC NETWORKS

F. P. KELLY

CAMBRIDGE
UNIVERSITY PRESS

CAMBRIDGE
UNIVERSITY PRESS

University Printing House, Cambridge CB2 8BS, United Kingdom

One Liberty Plaza, 20th Floor, New York, NY 10006, USA

477 Williamstown Road, Port Melbourne, VIC 3207, Australia

314-321, 3rd Floor, Plot 3, Splendor Forum, Jasola District Centre, New Delhi - 110025, India

103 Penang Road, #05-06/07, Visioncrest Commercial, Singapore 238467

Cambridge University Press is part of the University of Cambridge.

It furthers the University's mission by disseminating knowledge in the pursuit of education, learning and research at the highest international levels of excellence.

www.cambridge.org
Information on this title: www.cambridge.org/9781107401150

Original publisher: published 1979
Republished with a new preface by Cambridge University Press 2011

A catalogue record for this publication is available from the British Library

ISBN 978-1-107-40115-0 Paperback

Preface to the reissued edition

Networks have long been natural models in the physical, biological and social sciences, and our modern interconnected society relies increasingly upon our communication and transport networks. In response, there has been an explosion of interest in networks as objects worthy of study in their own right. I am pleasantly surprised by the continuing interest in this book, written to give an elementary introduction to the behaviour in equilibrium of stochastic networks, and I am grateful to Cambridge University Press for their encouragement to reissue the volume.

Frank Kelly

Cambridge, Christmas 2010

Preface

The main topic of this book is the study of the behaviour in equilibrium of vector stochastic processes, or stochastic networks. Such processes have a wide range of applications: to give some examples, the components of the vector may represent queue sizes in a queueing network, gene frequencies in a population, or the condition of fruit trees in an orchard. When a stochastic network is reversible its analysis is greatly simplified, and the first chapter is devoted to a discussion of the concept of reversibility. Two themes emerge from the remainder of the book: first, the various uses of reversibility, in the study of the output from a queue, the flow of current in a conductor, the age of an allele, or the equilibrium distribution of a polymerization process; second, the extent to which the assumption of reversibility can be relaxed without destroying the associated tractability.

The main prerequisite is an understanding of Markov processes at about the level of Feller's *Introduction to Probability Theory and Its Applications*, Volume I. In Section 1.1 the necessary material is very briefly reviewed, primarily to establish terminology and notation.

For their comments and advice I am indebted to many people, particularly Dave Aldous, Andrew Barbour, Dieter Koenig, Rolf Schassberger, and Geoff Watterson. I am especially grateful to Peter Whittle, whose lectures on reversibility first interested me in the subject and without whose encouragement the book would not have been written. Finally, my thanks go to Jackie Kelly for computing the graphs in the book and to Angie Ashton for typing the final draft.

Cambridge, Christmas 1978

FRANK KELLY

Contents

CHAPTER 1

Markov Processes and Reversibility

In this chapter the concept of reversibility is introduced and explored, and some simple stochastic models are described. The rest of the book will be devoted to generalizations of these simple models.

The first section reviews some aspects of the theory of Markov processes which will be required in the sequel.

1.1 PRELIMINARIES ON MARKOV PROCESSES

Let $X(t)$ be a stochastic process taking values in a countable state space $\mathcal{S}$ for $t \in \mathcal{T}$. Thus $(X(t_1), X(t_2), \ldots, X(t_n))$ has a known distribution for $t_1, t_2, \ldots, t_n \in \mathcal{T}$. For a discrete time stochastic process $\mathcal{T}$ will be the integers $\mathbb{Z}$ while for a continuous time stochastic process $\mathcal{T}$ will be the real line $\mathbb{R}$. These are the only possibilities we shall consider.

If $(X(t_1), X(t_2), \ldots, X(t_n))$ has the same distribution as $(X(t_1+\tau), X(t_2+\tau), \ldots, X(t_n+\tau))$ for all $t_1, t_2, \ldots, t_n, \tau \in \mathcal{T}$ then the stochastic process $X(t)$ is *stationary*.

The stochastic process $X(t)$ is a *Markov* process if for $t_1 < t_2 < \cdots < t_n < t_{n+1}$ the joint distribution of $(X(t_1), X(t_2), \ldots, X(t_n), X(t_{n+1}))$ is such that

$$P(X(t_{n+1}) = j_{n+1} \mid X(t_1) = j_1, X(t_2) = j_2, \ldots, X(t_n) = j_n)$$
$$= P(X(t_{n+1}) = j_{n+1} \mid X(t_n) = j_n)$$

whenever the conditioning event $(X(t_1) = j_1, X(t_2) = j_2, \ldots, X(t_n) = j_n)$ has positive probability. Where no confusion can arise we shall use an abbreviated notation in which the above equation becomes

$$P(j_{n+1} \mid j_1, j_2, \ldots, j_n) = P(j_{n+1} \mid j_n)$$

Thus for a Markov process the state of the process at a given time contains all the information about the past evolution of the process which is of use in predicting its future behaviour. This is the usual definition of a Markov process. An alternative equivalent definition is the following. The stochastic process $X(t)$ is a Markov process if for $t_1 < t_2 < \cdots < t_p < \cdots < t_m$, conditional on $X(t_p) = j_p$ (the present), $(X(t_1), X(t_2), \ldots, X(t_{p-1}))$ (the past) and $(X(t_{p+1}), X(t_{p+2}), \ldots, X(t_m))$ (the future) are independent (Exercise 1.1.2).

A Markov process is *time homogeneous* if $P(X(t+\tau) = k \mid X(t) = j)$ does not depend upon t, and is *irreducible* if every state in $\mathcal{S}$ can be reached from every other state. For a time homogeneous discrete time Markov process

$$p(j, k) = P(X(t+1) = k \mid X(t) = j)$$

is called the *transition probability* from state j to state k. Note that

$$\sum_{k \in \mathcal{S}} p(j, k) = 1 \qquad j \in \mathcal{S}$$

A discrete time Markov process is *periodic* if there exists an integer $\delta > 1$ such that $P(X(t+\tau) = j \mid X(t) = j) = 0$ unless τ is divisible by δ; otherwise the process is *aperiodic*. Throughout this work we shall assume that any discrete time Markov process with which we deal is time homogeneous and irreducible; we shall often additionally assume it is aperiodic. Consider then a process satisfying all these assumptions. Such a process may possess an *equilibrium distribution*, that is a collection of positive numbers $\pi(j)$, $j \in \mathcal{S}$, summing to unity that satisfy the equilibrium equations

$$\pi(j) = \sum_{k \in \mathcal{S}} \pi(k) p(k, j) \qquad j \in \mathcal{S} \tag{1.1}$$

If we can find a collection of positive numbers satisfying equations (1.1) whose sum is finite, then the collection can be normalized to produce an equilibrium distribution. When an equilibrium distribution exists it is unique and

$$\lim_{t \to \infty} P(X(t) = k \mid X(0) = j) = \pi(k) \tag{1.2}$$

so that π is the limiting distribution. Also, the proportion of time the process spends in state k during the period $[0, t]$ converges to $\pi(k)$ as $t \to \infty$, that is the process is ergodic. Further, if $P(X(0) = j) = \pi(j)$, $j \in \mathcal{S}$, then $P(X(t) = j) = \pi(j)$, $j \in \mathcal{S}$, for all $t \in \mathbb{Z}$, so that π is the stationary distribution. If an equilibrium distribution does not exist then

$$\lim_{t \to \infty} P(X(t) = k) = 0 \qquad k \in \mathcal{S}$$

and the process cannot be stationary. An equilibrium distribution will not exist if we can find a collection of positive numbers satisfying equations (1.1) whose sum is infinite. An equilibrium distribution will always exist when $\mathcal{S}$ is finite. All of the above remains true for periodic processes, except for the relation (1.2).

It is possible to construct continuous time Markov processes which exhibit extremely strange behaviour. These will be excluded; throughout this work we shall assume that any continuous time Markov process with which we deal is not only time homogeneous and irreducible but also remains in each state for a positive length of time and is incapable of passing through an infinite number of states in a finite time. Define the *transition rate* from state j to state k to be

$$q(j, k) = \lim_{\tau \to 0} \frac{P(X(t+\tau) = k \mid X(t) = j)}{\tau} \qquad j \neq k$$

It will be convenient to let $q(j, j) = 0$. For continuous time processes the equilibrium equations are

$$\pi(j) \sum_{k \in \mathcal{S}} q(j, k) = \sum_{k \in \mathcal{S}} \pi(k) q(k, j) \qquad j \in \mathcal{S} \tag{1.3}$$

and an equilibrium distribution is a collection of positive numbers $\pi(j)$, $j \in \mathcal{S}$, summing to unity which satisfy the equilibrium equations. As for discrete time processes an equilibrium distribution is unique if it exists and is then both the limiting and the stationary distribution. Further, the process is ergodic. If one does not exist then

$$\lim_{t \to \infty} P(X(t) = k) = 0 \qquad k \in \mathcal{S}$$

An equilibrium distribution will not exist if we can find a collection of positive numbers satisfying equations (1.3) whose sum is infinite. When $\mathcal{S}$ is finite an equilibrium distribution will always exist.

A discrete time Markov process is sometimes called a Markov chain. We shall use this terminology so that from now on when we refer to a Markov process it will be a continuous time process. We shall often refer to a stationary Markov chain or process as being in equilibrium.

A Markov process remains in state j for a length of time which is exponentially distributed with parameter

$$q(j) = \sum_{k \in \mathcal{S}} q(j, k)$$

When it leaves state j it moves to state k with probability

$$p(j, k) = \frac{q(j, k)}{q(j)} \tag{1.4}$$

There is thus a natural way to associate a Markov chain $X^J(t)$ with a Markov process $X(t)$. Let $X^J(0)$ be $X(0)$, let $X^J(1)$ be the next state the Markov process $X(t)$ moves to after time $t = 0$, let $X^J(2)$ be the next state after that, and so on. The Markov chain $X^J(t)$ is called the *jump chain* of the Markov process $X(t)$, and its transition probabilities are given by the relation (1.4). The equilibrium distribution of a jump chain will in general be different from that of the Markov process generating it (Exercise 1.1.5), essentially because the jump chain ignores the length of time the process remains in each state. The initial distribution at time $t=0$ of the jump chain of a stationary Markov process will be the equilibrium distribution of the Markov process, and thus the jump chain will not in general be stationary.

Exercises 1.1

1. Let $Z(t)$, $t\in\mathbb{Z}$, be a sequence of independent identically distributed random variables with $P(Z(t)=0)=P(Z(t)=1)=\frac{1}{2}$. Define the stochastic process $X(t)$ with $\mathcal{S}=\{0,1,2,\ldots,6\}$ and $\mathcal{T}=\mathbb{Z}$ by $X(t)=Z(t-1)+2Z(t)+3Z(t+1)$.
   (a) Determine $P(X(0)=1,\quad X(1)=3,\quad X(2)=2)$ and $P(X(1)=3, X(2)=2)$.
   (b) Determine $P(X(2)=2 \mid X(0)=1,\quad X(1)=3)$ and $P(X(2)=2 \mid X(1)=3)$. Deduce that the process $X(t)$ is not Markov.
2. Establish the equivalence of the following statements:
   (i) For all $t_1<t_2<\cdots<t_n<t_{n+1}$,
   $$P(j_{n+1} \mid j_1, j_2, \ldots, j_n)=P(j_{n+1} \mid j_n)$$
   (ii) For all $t_1<t_2<\cdots<t_p<\cdots<t_m$,
   $$P(j_1, j_2, \ldots, j_{p-1}, j_{p+1}, j_{p+2}, \ldots, j_m \mid j_p)$$
   $$=P(j_1, j_2, \ldots, j_{p-1} \mid j_p)P(j_{p+1}, j_{p+2}, \ldots, j_m \mid j_p)$$
3. If a Markov process has an equilibrium distribution show that the convergence to it expressed in the relation (1.2) is uniform over states $k\in\mathcal{S}$.
4. Consider the Markov process with state space $\mathcal{S}=\{0,1,2,\ldots\}$ and with transition rates
   $$q(j,k)=\begin{cases} a^j & k=j+1\\ b & k=0\\ 0 & \text{otherwise}\end{cases}$$
   If $a>1$ this process is capable of passing through an infinite number of states in finite time. Find the equilibrium distribution when $a\leq 1$ and $b>0$. Observe that one does not exist when $0<a\leq 1$ and $b=0$.
5. It is possible for a Markov process to possess an equilibrium distribution and for its jump chain not to, and vice versa. Show that if a Markov process has equilibrium distribution $\pi(j)$, $j\in\mathcal{S}$, then its jump chain has an equilibrium distribution if and only if
   $$B^{-1}=\sum_{j\in\mathcal{S}}\pi(j)q(j)$$

is finite, in which case the equilibrium distribution of the jump chain is

$$\pi^J(j) = B\pi(j)q(j).$$

Observe that if $q(j)$ does not depend upon j, so that the points in time at which jumps take place form a Poisson process, then the jump chain and the process have the same equilibrium distribution.

1.2 REVERSIBILITY

Some stochastic processes have the property that when the direction of time is reversed the behaviour of the process remains the same. Speaking intuitively, if we take a film of such a process and then run the film backwards the resulting process will be statistically indistinguishable from the original process. This property is described formally in the following definition.

Definition

A stochastic process $X(t)$ is *reversible* if $(X(t_1), X(t_2), \ldots, X(t_n))$ has the same distribution as $(X(\tau - t_1), X(\tau - t_2), \ldots, X(\tau - t_n))$ for all $t_1, t_2, \ldots, t_n, \tau \in \mathcal{T}$.

In the next section we shall give examples of reversible processes and in later sections we shall discuss some of the less obvious consequences of the above definition; but first let us derive some of the basic properties of reversible processes.

Lemma 1.1. *A reversible process is stationary.*

Proof. Since $X(t)$ is reversible both $(X(t_1), X(t_2), \ldots, X(t_n))$ and $(X(t_1+\tau), X(t_2+\tau), \ldots, X(t_n+\tau))$ have the same distribution as $(X(-t_1), X(-t_2), \ldots, X(-t_n))$. Hence $X(t)$ is stationary.

For a stationary Markov chain or process there exist simple necessary and sufficient conditions for reversibility given in terms of the equilibrium distribution and the transition probabilities or rates. These conditions are obtained in Theorems 1.2 and 1.3 and are called the detailed balance conditions; they should be contrasted with the equilibrium equations, which are sometimes called the full balance conditions.

Theorem 1.2. *A stationary Markov chain is reversible if and only if there exists a collection of positive numbers $\pi(j)$, $j \in \mathcal{S}$, summing to unity that satisfy the detailed balance conditions*

$$\pi(j)p(j, k) = \pi(k)p(k, j) \qquad j, k \in \mathcal{S} \tag{1.5}$$

When there exists such a collection $\pi(j)$, $j \in \mathscr{S}$, it is the equilibrium distribution of the process.

Proof. First suppose that the process is reversible. Since the process is stationary $P(X(t)=j)$ does not depend upon t. Let $\pi(j)=P(X(t)=j)$; thus $\pi(j)$, $j \in \mathscr{S}$, is a collection of positive numbers summing to unity. Since the process is reversible

$$P(X(t)=j, X(t+1)=k)=P(X(t)=k, X(t+1)=j)$$

and so

$$\pi(j)p(j,k)=\pi(k)p(k,j)$$

Conversely, suppose there exists a collection of positive numbers $\pi(j)$, $j \in \mathscr{S}$, summing to unity satisfying the detailed balance conditions. Summing equations (1.5) over k we obtain

$$\pi(j)\sum_{k\in\mathscr{S}} p(j,k)=\sum_{k\in\mathscr{S}}\pi(k)p(k,j) \qquad j\in\mathscr{S}$$

which reduce to the equilibrium equations (1.1). Hence the collection $\pi(j)$, $j \in \mathscr{S}$, is the equilibrium distribution of the process. Consider now a sequence of states $j_0, j_1, \ldots, j_m$. Then

$$P(X(t)=j_0, X(t+1)=j_1, \ldots, X(t+m)=j_m)$$
$$=\pi(j_0)p(j_0,j_1)p(j_1,j_2)\cdots p(j_{m-1},j_m)$$

and

$$P(X(t')=j_m, X(t'+1)=j_{m-1}, \ldots, X(t'+m)=j_0)$$
$$=\pi(j_m)p(j_m,j_{m-1})p(j_{m-1},j_{m-2})\cdots p(j_1,j_0)$$

But the detailed balance conditions (1.5) imply that the right-hand sides of the last two identities are equal. Hence, letting $\tau=t+t'+m$, $(X(t), X(t+1), \ldots, X(t+m))$ has the same distribution as $(X(\tau-t), X(\tau-t-1), \ldots, X(\tau-t-m))$, and from this we can deduce that $(X(t_1), X(t_2), \ldots, X(t_n))$ has the same distribution as $(X(\tau-t_1), X(\tau-t_2), \ldots, X(\tau-t_n))$ for all $t_1, t_2, \ldots, t_n$, $\tau \in \mathbb{Z}$.

The detailed balance conditions (1.5) imply that if $p(j,k)$ is positive then so is $p(k,j)$. Less obvious, but interesting, consequences for the matrix of transition probabilities are contained in Exercises 1.2.4 and 1.2.5.

Theorem 1.2 has a direct analogue for continuous time processes.

Theorem 1.3. *A stationary Markov process is reversible if and only if there exists a collection of positive numbers $\pi(j)$, $j \in \mathscr{S}$, summing to unity that satisfy* *the detailed balance conditions*

$$\pi(j)q(j,k)=\pi(k)q(k,j) \qquad j,k\in\mathcal{S} \tag{1.6}$$

When there exists such a collection $\pi(j)$, $j\in\mathcal{S}$, it is the equilibrium distribution of the process.

Proof. First suppose the process is reversible, and let $\pi(j)=P(X(t)=j)$. Then

$$P(X(t)=j, X(t+\tau)=k)=P(X(t)=k, X(t+\tau)=j)$$

and so

$$\pi(j)\frac{P(X(t+\tau)=k \mid X(t)=j)}{\tau}=\pi(k)\frac{P(X(t+\tau)=j \mid X(t)=k)}{\tau}$$

Letting $\tau\to 0$ we obtain the relation (1.6).

Conversely, suppose there exists the collection $\pi(j)$, $j\in\mathcal{S}$, satisfying the detailed balance conditions. Summing equations (1.6) over k gives the equilibrium equations (1.3), and hence the collection $\pi(j)$, $j\in\mathcal{S}$, is the equilibrium distribution. Consider now the behaviour of the process $X(t)$ for $t\in[-T,T]$. The process may start at time $t=-T$ in state j_1 and remain in this state for a period h_1 before jumping to state j_2. Suppose it now remains in state j_2 for a period h_2 before jumping to state j_3, and so on, until it arrives in state j_m where it remains until time $t=T$, a period of h_m, say. Now the probability density of the random variable h_1 is

$$q(j_1)e^{-q(j_1)h_1}$$

and the probability that j_2 is the next state after j_1 is

$$\frac{q(j_1,j_2)}{q(j_1)}$$

Similarly, we can calculate the density of h_2 and the probability that j_3 is the next state after j_2, and so on. The probability that the process remains in state j_m for a period of at least h_m is

$$e^{-q(j_m)h_m}$$

Thus the probability density of the behaviour described is

$$\pi(j_1)e^{-q(j_1)h_1}q(j_1,j_2)e^{-q(j_2)h_2}q(j_2,j_3)e^{-q(j_3)h_3}\cdots q(j_{m-1},j_m)e^{-q(j_m)h_m} \tag{1.7}$$

This is a density with respect to $h_1, h_2, \ldots, h_m$. To obtain a probability it must be integrated over a region of values for $h_1, h_2, \ldots, h_m$ satisfying the constraint $h_1+h_2+\cdots+h_m=2T$. Now the relation (1.6) implies that

$$\pi(j_1)q(j_1,j_2)q(j_2,j_3)\cdots q(j_{m-1},j_m)=\pi(j_m)q(j_m,j_{m-1})\cdots q(j_3,j_2)q(j_2,j_1)$$

and hence that expression (1.7) is equal to the probability density that the process starts at time $t=-T$ in state j_m, that it remains in this state for a period h_m before jumping to state j_{m-1}, and so on, until it arrives in state j_1 which it remains in until time $t=T$, a period of h_1. Thus the probabilistic behaviour of $X(t)$ is precisely the same as that of $X(-t)$ on the interval $[-T, T]$. Thus $(X(t_1), X(t_2), \ldots, X(t_m))$ has the same distribution as $(X(-t_1), X(-t_2), \ldots, X(-t_m))$, but this has the same distribution as $(X(\tau-t_1), X(\tau-t_2), \ldots, X(\tau-t_m))$ because $X(t)$ is stationary; and so the theorem is proved.

A collection of positive numbers satisfying the detailed balance conditions whose sum is finite can of course be normalized to produce an equilibrium distribution. Lemma 1.1 shows that a Markov process which is not stationary is not reversible, even if the detailed balance condition can be satisfied.

The term $\pi(j)q(j, k)$ is called the *probability flux* from state j to state k; in equilibrium the probability that in the interval $(t, t+\delta t)$ the process jumps from state j to state k is $\pi(j)q(j, k)\,\delta t+o(\delta t)$. The detailed balance condition (1.6) requires that the probability flux from state j to k should equal that from state k to j. The full balance condition (1.3) requires that the probability flux out of state j should equal that into state j. These relationships can perhaps be more easily visualized if we associate a graph G with the Markov process as follows: let the set of vertices of the graph be $\mathcal{S}$, the set of states, and let there be an edge joining vertices j and k if either $q(j, k)$ or $q(k, j)$ is positive. Thus the Markov process can be regarded as a random walk on the graph G. Note that the assumed irreducibility of the process implies that the graph is connected. Define a cut to be a division of $\mathcal{S}$ into complementary sets $\mathcal{A}$ and $\mathcal{S}-\mathcal{A}$.

Lemma 1.4. *For a stationary Markov process the probability flux each way across a cut balances. That is for any $\mathcal{A}\subset\mathcal{S}$,*

$$\sum_{j\in\mathcal{A}}\sum_{k\in\mathcal{S}-\mathcal{A}}\pi(j)q(j,k)=\sum_{j\in\mathcal{A}}\sum_{k\in\mathcal{S}-\mathcal{A}}\pi(k)q(k,j) \tag{1.8}$$

Proof. Summing the full balance condition (1.3) over $j\in\mathcal{A}$ gives

$$\sum_{j\in\mathcal{A}}\sum_{k\in\mathcal{S}}\pi(j)q(j,k)=\sum_{j\in\mathcal{A}}\sum_{k\in\mathcal{S}}\pi(k)q(k,j)$$

The result follows by subtracting the identity

$$\sum_{j\in\mathcal{A}}\sum_{k\in\mathcal{A}}\pi(j)q(j,k)=\sum_{j\in\mathcal{A}}\sum_{k\in\mathcal{A}}\pi(k)q(k,j)$$

Note that if $\mathcal{A}=\{j\}$ then equations (1.8) reduce to the equilibrium equations (1.3).

Lemma 1.5. *If the graph G associated with a stationary Markov process is a tree, then the process is reversible.*

Proof. If j and k are not linked by an edge of the graph G the detailed balance condition (1.6) is satisfied trivially. If j and k are linked by an edge then removal of this edge cuts the graph G into two unconnected components, since G is a tree. Thus Lemma 1.4 shows that the detailed balance condition is satisfied.

Lemma 1.5 gives a sufficient condition for a process to be reversible but, as we shall see later, it is by no means necessary.

It can be shown that the number of transitions from state j to k per unit time calculated over the period $(0, t)$ converges to $\pi(j)q(j, k)$ as $t \to \infty$. This fact provides an alternative proof of Lemmas 1.4 and 1.5 since the number of transitions each way across a cut in the period $(0, t)$ cannot differ by more than one.

Lemmas 1.4 and 1.5 have obvious counterparts for Markov chains.

Exercises 1.2

1. Consider the stationary Markov process $X(t)$ with $\mathcal{S} = \{1, 2\}$, $q(1, 2) = 1$, $q(2, 1) = \frac{1}{2}$. Show that $X(t)$ is reversible. Observe that a film of the process, run either forwards or backwards, will show the process alternating between states with the periods in states 1 and 2 having means 1 and 2 respectively. There is a minor difficulty here which should be pointed out. Suppose the process jumps from state 1 to 2 at time t_0. Is $X(t_0) = 1$ or 2? The usual convention is that if $X(t)$ jumps at time t_0 then $X(t_0)$ is taken to be the new state, so that the process is right continuous. The difficulty is that if the film run forwards is right continuous then the film run backwards will be left continuous. The difficulty is avoided if we adopt the convention that $X(t_0)$ is equally likely to be the old or the new state. Such fine differences would of course be hard to detect (the finite dimensional distributions do not manage it), and will not concern us from now on. When, later, we speak of the instant in time just preceding (respectively, following) a transition we shall be implicitly appealing to the left (respectively, right) continuous version of the process.
2. Show that the stochastic process $X(t)$ defined in Exercise 1.1.1 is not reversible.
3. Suppose that the points $s_i \in \mathbb{R}$, $i = \ldots, -1, 0, 1, 2, \ldots$, form a Poisson process and define

$$X(t) = \sum_{i=-\infty}^{+\infty} a(s_i - t)$$

Show that $X(t)$ is reversible if

$$a(s)=\begin{cases}1 & -1<s\le 0\\ 0 & \text{otherwise}\end{cases}$$

and is not reversible if

$$a(s)=\begin{cases}2 & -1<s\le 0\\ 1 & -2<s\le -1\\ 0 & \text{otherwise}\end{cases}$$

4. Show that a stationary Markov chain is reversible if and only if the matrix of transition probabilities can be written as the product of a symmetric and a diagonal matrix.
5. Show that the matrix of transition probabilities of a reversible Markov chain can be written in the form $D^{-1}AD$ where D is diagonal and A symmetric. Deduce that it has real eigenvalues (the converse is false as the next exercise shows).
6. Consider a stationary Markov chain with the following matrix of transition probabilities:

$$\begin{pmatrix}0 & 1 & 0\\ \frac{3}{4} & 0 & \frac{1}{4}\\ 1 & 0 & 0\end{pmatrix}$$

   Show that the process is not reversible even though the matrix has real eigenvalues.
7. Suppose a Markov process and its jump chain both possess equilibrium distributions. Observe that the equilibrium probability that the jump chain is in state j, found in Exercise 1.1.5, is proportional to the probability flux out of, or, equivalently, the probability flux into, state j in the Markov process. Show that the transition rates of the Markov process satisfy the detailed balance conditions if and only if the transition probabilities of the jump chain do.
8. If $X_1(t)$ and $X_2(t)$ are independent reversible Markov processes show that $(X_1(t), X_2(t))$ is a reversible Markov process.
9. If $X(t)$ is a reversible stochastic process show that so is $Y(t)=f[X(t)]$ for any function f.

1.3 BIRTH AND DEATH PROCESSES

The simplest examples of reversible processes are provided by Markov processes for which the state space $\mathcal{S}$ is $\{0, 1, 2, \ldots, K\}$, with K possibly infinite, and $q(j, k)=0$ unless $|j-k|=1$. These are called birth and death processes, since the only possible transitions from state j are to $j-1$ (a death) or $j+1$ (a birth). A stationary birth and death process is reversible, by Lemma 1.5. The detailed balance condition states that the equilibrium distribution of a stationary birth and death process satisfies

$$\pi(j)q(j, j-1) = \pi(j-1)q(j-1, j)$$

and hence is given by

$$\pi(j) = \pi(0)\prod_{r=1}^{j}\frac{q(r-1, r)}{q(r, r-1)} \tag{1.9}$$

where $\pi(0)$ must be chosen so that $\pi(j)$, $j = 0, 1, 2, \ldots$, sum to unity. If $\pi(0)$ cannot be so chosen then the process does not possess an equilibrium distribution and cannot be stationary. We will now discuss some simple examples of birth and death processes.

The simple queue. Suppose that the stream of customers arriving at a queue (the arrival process) forms a Poisson process of rate ν. Suppose further that there is a single server and that customers' service times are independent of each other and of the arrival process and are exponentially distributed with mean μ^{-1}. Such a queue is called simple or *M/M/*1, the *M*'s indicating the memoryless (exponential) character of the interarrival and service times and the final digit indicating the number of servers. Let $n(t)$ be the number of customers in the queue at time t, including the customer being served. Then it follows from our description of the queue that $n(t)$ is a birth and death process with transition rates

$$q(j, j-1) = \mu \qquad j = 1, 2, \ldots$$
$$q(j, j+1) = \nu \qquad j = 0, 1, \ldots$$

If the arrival rate ν is less than the service rate μ the process has an equilibrium distribution which is, from equation (1.9),

$$\pi(j) = \left(1 - \frac{\nu}{\mu}\right)\left(\frac{\nu}{\mu}\right)^j \tag{1.10}$$

Thus in equilibrium the number in the queue has a geometric distribution with mean $\nu/(\mu-\nu)$.

This result can be used to obtain another distribution of interest, the distribution of the waiting time of a customer. We shall define waiting time to include service time, so that it is the period between a customer's arrival at and departure from the queue. Consider now a typical customer arriving at the queue and let W be his waiting time. Assume for the moment that the probability he finds j customers already present in the queue is $\pi(j)$. With the queue discipline first come first served,

$$P(W \le w) = \sum_{j=0}^{\infty}\pi(j)P\left(\sum_{r=1}^{j+1} S_r \le w\right) \tag{1.11}$$

where $S_1, S_2, \ldots$ are independent exponentially distributed random variables with mean μ^{-1}. After some reduction (Exercise 1.3.1) this shows that W is exponentially distributed with mean $(\mu-\nu)^{-1}$.

Is it valid to assume that when a typical customer arrives at the queue he finds it in equilibrium? This assumption can be made when the arrival process is Poisson, although we must be careful about the interpretation of a typical customer. If we observe a customer arriving at time t_0 and we know nothing other than this about arrival times or about the state of the queue, then we shall call this customer typical. When the arrival process is Poisson the interval between t_0 and the preceding arrival has an exponential distribution, and indeed the arrival process up until time t_0 has the same probabilistic description as it would have if t_0 were just a fixed instant in time. Hence the customer arriving at time t_0 finds the queue in equilibrium. (The concept of a typical customer is investigated further in Exercise 1.3.7.)

There is an alternative approach to this result which is of greater generality and will be of use later. The probability that in the interval $(t_0, t_0+\delta t)$ a single customer arrives and finds j customers already present in the queue is

$$\pi(j)q(j, j+1)\,\delta t + o(\delta t)$$

The probability that in the interval $(t_0, t_0+\delta t)$ a single customer arrives is

$$\sum_{j=0}^{\infty} \pi(j)q(j, j+1)\,\delta t + o(\delta t)$$

Thus given that a single customer arrives in the interval $(t_0, t_0+\delta t)$ the conditional probability that he finds j customers already present in the queue is

$$\frac{\pi(j)q(j, j+1)\,\delta t + o(\delta t)}{\sum_{j=0}^{\infty} \pi(j)q(j, j+1)\,\delta t + o(\delta t)}$$

As $\delta t \to 0$ this conditional probability tends to the ratio

$$\frac{\pi(j)q(j, j+1)}{\sum_{j=0}^{\infty} \pi(j)q(j, j+1)}$$

The numerator is the probability flux that a customer arrives to find j customers already present in the queue, and the denominator is the probability flux that a customer arrives. Thus this ratio is also the limit as $t \to \infty$ of the proportion of arrivals in the period $(0, t)$ who find j customers already present in the queue. Since $q(j, j+1) = \nu$ the ratio is simply $\pi(j)$.

The above approach is of use whenever a stochastic process is observed at just those points in time marked by some special event. Exercises 1.1.5, 1.3.6, and 1.3.9 provide further examples.

For a simple queue the mean number in the queue $E(n)$, the mean waiting time of a customer $E(W)$, and the mean time between successive arrivals ν^{-1} are related by the identity

$$E(n) = \nu E(W). \tag{1.12}$$

This, Little's result, holds for much more general systems—the arrival process need not be Poisson, service times need not be independent, and indeed the system may bear little resemblance to a queue at all. It has the nature of an accounting identity; we can count time spent in a system either by adding it up over the customers who pass through the system or by integrating the number in the system over time. We shall not prove Little's result, although we shall occasionally use it. For our purposes it will be enough to record that equation (1.12) holds whenever there is a stationary Markov process $X(t)$ such that the number in the system at time t, $n(t)$, is a function of $X(t)$. The expectation $E(W)$ can be regarded as the mean time spent in the system by a typical customer or as the limit as $m \to \infty$ of the average time spent in the system by the first m customers to enter the system after time t. Similarly, ν can be regarded as the reciprocal of the mean interarrival period preceding the arrival of a typical customer or as the limit as $t \to \infty$ of the number of customers to arrive per unit time calculated over the period $(0, t)$. In equilibrium the probability flux that a customer arrives is ν. When the arrival process is not Poisson we shall call ν the mean arrival rate.

A telephone exchange. Suppose that calls are initiated at points in time which form a Poisson process of rate ν, but that the exchange has only K lines so that a call initiated when K calls are already in progress is lost. Further suppose that calls which are connected last for lengths of time which are independent and exponentially distributed with mean μ^{-1}. Then the number of calls in progress at time t is a birth and death process with transition rates

$$q(j, j-1) = j\mu \qquad j = 1, 2, \ldots, K$$

$$q(j, j+1) = \nu \qquad j = 0, 1, \ldots, K-1$$

The equilibrium distribution over the state space $\mathcal{S} = \{0, 1, \ldots, K\}$ is

$$\pi(j) = \pi(0)\frac{1}{j!}\left(\frac{\nu}{\mu}\right)^j$$

Thus in equilibrium the number of calls in progress has a truncated Poisson distribution.

The probability that a typical call will be lost is

$$\pi(K) = \frac{(1/K!)(\nu/\mu)^K}{\sum_{j=0}^{K}(1/j!)(\nu/\mu)^j} \tag{1.13}$$

This, Erlang's formula, also gives the limiting proportion of calls lost.

The simple birth, death, and immigration process. Suppose that individuals immigrate at rate ν, give birth to additional individuals at rate λ, and die at rate μ so that

$$q(j, j-1) = j\mu \qquad j = 1, 2, \ldots$$

$$q(j, j+1) = \nu + j\lambda \qquad j = 0, 1, \ldots$$

These transition rates correspond to the assumptions that the lifetimes of individuals are independent of each other and of the immigration process and that during an individual's lifetime the points in time at which it gives birth form a Poisson process independent of other lifetimes and of the immigration process. It is often tedious to specify precisely the assumptions underlying a model; where the assumptions are clear from the context or from the structure of a Markov process we shall often fail to list them. It follows from equation (1.9) that when $\lambda < \mu$ the equilibrium distribution for the number of individuals alive is

$$\pi(j) = \left(1 - \frac{\lambda}{\mu}\right)^{\nu/\lambda} \binom{\frac{\nu}{\lambda} + j - 1}{j} \left(\frac{\lambda}{\mu}\right)^{j} \qquad (1.14)$$

where

$$\binom{x}{r} = \frac{x(x-1)\cdots(x-r+1)}{r(r-1)\cdots 1}$$

This distribution is an example of the negative binomial distribution; its mean is $\nu/(\mu - \lambda)$ and its variance is $\nu\mu/(\mu - \lambda)^2$. When $\lambda = \nu$ it reduces to the geometric distribution (1.10).

Exercises 1.3

1. Relation (1.11) shows that W is the sum of $j+1$ independent exponentially distributed random variables, where j itself is a random variable with a geometric distribution. By considering the Markov process with three states and transition rates $q(1,2) = q(2,1) = \nu$, $q(1,3) = q(2,3) = \mu - \nu$, show that W is exponentially distributed with mean $(\mu - \nu)^{-1}$.
2. Suppose the simple queue described above is amended by the requirement that any customer who arrives when there are K customers already present must leave immediately without service. Show that in equilibrium the probability that this amended queue contains n customers is just the conditional probability that the simple queue contains n customers given that it contains not more than K customers.
3. Show that for an $M/M/s$ queue the number in the queue is a birth and

death process whose equilibrium distribution is determined by

$$\pi(j)=\pi(0)\left(\frac{\nu}{\mu}\right)^j\frac{1}{j!} \qquad j=1,2,\ldots,s$$

$$\pi(j)=\pi(s)\left(\frac{\nu}{s\mu}\right)^{j-s} \qquad j=s+1,s+2,\ldots$$

provided $\nu<s\mu$. The ratio $\rho=\nu/s\mu$ is called the traffic intensity. Show that if a typical customer arrives to find all the servers busy then, with the queue discipline first come first served, his queueing time (the period of time until his service commences) is exponentially distributed with mean $(s\mu-\nu)^{-1}$.

4. Suppose the number of customers in an $M/M/1$ queue is observed at those instants in time at which a customer is about to arrive. Show that the resulting discrete time process is a Markov chain with transition probabilities

$$p(j,k)=\frac{\nu}{\nu+\mu}\left(\frac{\mu}{\nu+\mu}\right)^{j-k+1} \qquad 0\le k\le j+1$$

Verify that the equilibrium distribution is given by the expression (1.10).

5. The Poisson assumption in the telephone exchange model may be adequate if the source population of subscribers is very large. If the source population is of finite size $M(>K)$, it may be more reasonable to let

$$q(j,j+1)=\lambda(M-j) \qquad j=0,1,\ldots,K-1$$

Show that the equilibrium distribution will then be given by

$$\pi(j)=\pi(0)\binom{M}{j}\left(\frac{\lambda}{\mu}\right)^j \qquad j=0,1,\ldots,K$$

6. Consider the finite source telephone exchange model of the previous exercise. Suppose the number of busy lines is observed at those instants in time at which a call is about to be initiated. Write down the transition probabilities of the resulting Markov chain. By considering the probability flux $\pi(j)q(j,j+1)$ that a call is initiated while j lines are busy show that the equilibrium distribution of the Markov chain is given by

$$\pi'(j)=\pi'(0)\binom{M-1}{j}\left(\frac{\lambda}{\mu}\right)^j \qquad j=0,1,2,\ldots,K$$

Comparing this distribution with the one obtained in the preceding exercise we see that the number of busy lines found by a subscriber when he attempts to make a call has the same distribution as the number of busy lines at a fixed instant in time in a system with one less subscriber.

7. Suppose that a typical customer arrives at an $M/M/1$ queue at time t_0. Show that the mth customer to arrive after time t_0 finds the queue in equilibrium, for $m = 1, 2, \ldots$. In contrast, the first customer to arrive after a fixed instant in time does not find the queue in equilibrium, since the interarrival period preceding his arrival is the sum of two exponential random variables. This customer is not typical: the way in which he has been chosen provides us with information about the time of previous arrivals. Show that the probability this customer finds j customers in the queue is

$$\begin{array}{ll} (1-\rho)(1+\rho) & j=0 \\ (1-\rho)\rho^{j+1} & j=1,2,\ldots \end{array}$$

where $\rho = \nu/\mu$.

8. A stack is a form of queue in which the server devotes his entire attention to the customer who last arrived at the queue. Thus when a customer arrives his service is started immediately, but is interrupted if another customer arrives before its completion. Suppose that customers are of I types, that the stream of customers of type i arriving at the queue forms a Poisson process of rate ν_i, and that the service times of these customers are exponentially distributed with parameter μ_i. Construct a Markov process to represent the queue and show that the graph associated with the process is a tree. Show that if

$$\rho = \sum_{i=1}^{I} \frac{\nu_i}{\mu_i} < 1$$

the equilibrium probability that there are n customers in the queue with the rth customer being of type $t(r)$ is

$$(1-\rho) \prod_{r=1}^{n} \frac{\nu_{t(r)}}{\mu_{t(r)}}$$

Deduce that in equilibrium the number of customers in the queue has the same distribution as for the simple queue with $\nu/\mu = \rho$.

9. Consider the points in time at which new individuals appear, either through immigration or birth, in the simple birth, death, and immigration process. Show that the mean time between such appearances is $(\mu - \lambda)/\nu\mu$ by using Little's result (1.12). Equivalently show that the mean appearance rate is $\nu\mu/(\mu - \lambda)$ by calculating the probability flux that a new individual appears. Show that when a new individual appears the number of individuals he finds already alive has a negative binomial distribution with mean $(\nu + \lambda)/(\mu - \lambda)$. Conditional on the new individual having been born show that the number of individuals he finds already alive, excluding his parent, has the same negative binomial distribution.

1.4 THE EHRENFEST MODEL

One particular example of a birth and death process is worth special study; it was introduced early in the century to help explain the apparent paradox between reversibility and the phenomenon of increasing entropy. The model can be described as follows. There are K particles distributed between two containers (Fig. 1.1). Particles behave independently and change container at rate λ. Thus $X(t)$, the number of particles in container 1 at time t, is a Markov process with transition rates

$$q(j, j-1) = j\lambda \qquad j = 1, 2, \ldots, K$$

$$q(j, j+1) = (K-j)\lambda \qquad j = 0, 1, \ldots, K-1$$

The equilibrium distribution can be deduced from equation (1.9) and is

$$\pi(j) = 2^{-K}\binom{K}{j}$$

The process in equilibrium is reversible and thus, assuming K is even,

$$P(X(t) = K, X(t+\tau) = \tfrac{1}{2}K) = P(X(t) = \tfrac{1}{2}K, X(t+\tau) = K) \qquad (1.15)$$

The equilibrium distribution shows that states which allocate particles fairly evenly between the two containers are much more likely than states which allocate most of the particles to one container. Hence the conditional probability $P(X(t+\tau) = \frac{1}{2}K \mid X(t) = K)$ is much greater than $P(X(t+\tau) = K \mid X(t) = \frac{1}{2}K)$. If the process starts with all the particles in one container then it is quite likely that after a period the particles will be shared evenly between the two containers. On the other hand, if the process starts with the particles shared evenly between the containers it is extremely unlikely that after a period the particles will all be in one container. The lack of symmetry exhibited by the conditional probabilities is quite compatible with reversibility. It is joint probabilities, such as those appearing in equation (1.15), which reversibility requires to be symmetric.

The asymmetry of the conditional probabilities, and more generally the phenomenon of increasing entropy, is a symptom of the approach to equilibrium of a system not initially in equilibrium. Consider a Markov process $X(t)$ with a finite state space. Let

$$u_j(t) = P(X(t) = j)$$

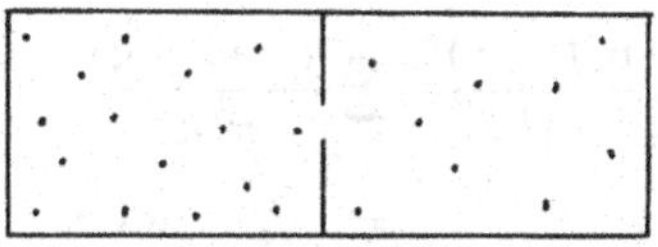

Fig. 1.1 The Ehrenfest model

and suppose that the initial distribution, $u_j(0)$, $j \in \mathcal{S}$, may not be the equilibrium distribution. Considering the possible events in the time interval $(t, t+\delta t)$ leads to the equation

$$u_j(t+\delta t) = \sum_{k\in\mathcal{S}} u_k(t)q(k,j)\,\delta t + u_j(t)\left(1 - \sum_{k\in\mathcal{S}} q(j,k)\,\delta t\right) + o(\delta t)$$

and hence the forward equations

$$\frac{\mathrm{d}}{\mathrm{d}t}u_j(t) = \sum_{k\in\mathcal{S}} (u_k(t)q(k,j) - u_j(t)q(j,k)) \quad j \in \mathcal{S} \tag{1.16}$$

The solution to these equations must satisfy the initial conditions at time $t=0$, and tends to the equilibrium distribution as $t \to \infty$. Now let

$$H(t) = \sum_{j\in\mathcal{S}} \pi(j)h\left(\frac{u_j(t)}{\pi(j)}\right)$$

where $h(x)$ is a strictly concave function. Thus $H(t)$ is a function of the distribution over states at time t, $u_j(t)$, $j \in \mathcal{S}$. If the initial distribution is the equilibrium distribution, then $H(t)$ takes a constant value. Otherwise $H(t)$ increases monotonically to this constant value, as the next theorem shows.

Theorem 1.6. *If the initial distribution is not the equilibrium distribution, then the function $H(t)$, $t > 0$, is strictly increasing.*

Proof. For fixed $\tau > 0$ let

$$p(j,k) = P(X(t+\tau) = k \mid X(t) = j)$$

Thus

$$u_k(t+\tau) = \sum_j u_j(t)p(j,k)$$

and

$$\pi(k) = \sum_j \pi(j)p(j,k)$$

Let

$$a(k,j) = \frac{\pi(j)p(j,k)}{\pi(k)} \tag{1.17}$$

Thus $a(k,j) > 0$, $\sum_j a(k,j) = 1$. Also

$$\frac{u_k(t+\tau)}{\pi(k)} = \sum_j \frac{u_j(t)p(j,k)}{\pi(k)}$$

$$= \sum_j a(k,j)\frac{u_j(t)}{\pi(j)} \tag{1.18}$$

Now since $h(x)$ is strictly concave

$$h\left(\sum_j a(k, j)x_j\right) > \sum_j a(k, j)h(x_j) \tag{1.19}$$

unless x_j, $j \in \mathcal{S}$, are all equal. Using successively relations (1.18), (1.19), and (1.17) we have that unless $u_j(t) = \pi(j)$, $j \in \mathcal{S}$,

$$\begin{aligned} H(t+\tau) &= \sum_k \pi(k)h\left(\frac{u_k(t+\tau)}{\pi(k)}\right) \\ &= \sum_k \pi(k)h\left(\sum_j a(k, j)\frac{u_j(t)}{\pi(j)}\right) \\ &> \sum_j \sum_k \pi(k)a(k, j)h\left(\frac{u_j(t)}{\pi(j)}\right) \\ &= \sum_j \sum_k \pi(j)p(j, k)h\left(\frac{u_j(t)}{\pi(j)}\right) \\ &= H(t) \end{aligned}$$

The theorem has a counterpart for Markov chains which is established in the same way.

An important special case of the theorem arises with the concave function $h(x) = -x \log x$. Then

$$H(t) = -\sum_j u_j(t) \log \frac{u_j(t)}{\pi(j)}$$

This quantity is called the statistical entropy, or the entropy of the distribution $u_j(t)$, $j \in \mathcal{S}$, with respect to the distribution $\pi(j)$, $j \in \mathcal{S}$.

The monotonic increase of the function $H(t)$ is a consequence of the convergence of the distribution $u_j(t)$, $j \in \mathcal{S}$, to the equilibrium distribution $\pi(j)$, $j \in \mathcal{S}$. It will occur whether or not the Markov process $X(t)$ has transition rates which satisfy the detailed balance conditions (1.6), provided only that the process is not in equilibrium. On the other hand, reversibility is essentially a property which a process in equilibrium may or may not possess, and in either case the function $H(t)$ is constant just because the process is in equilibrium. To take the example of the Ehrenfest model, there is no conflict between reversibility and the phenomenon of increasing entropy—reversibility is a property of the model in equilibrium and increasing entropy is a property of the approach to equilibrium.

If the transition rates of the Markov process $X(t)$ do satisfy the detailed balance conditions then there is an interesting alternative interpretation of the approach to equilibrium and of the function $H(t)$. In this case the forward equations (1.16) can be rewritten

$$\frac{d}{dt}u_j(t)=\sum_{k\in\mathcal{S}}\frac{1}{r(j,k)}\left(\frac{u_k(t)}{\pi(k)}-\frac{u_j(t)}{\pi(j)}\right)\qquad j\in\mathcal{S}\tag{1.20}$$

where the (possibly infinite) quantity $r(j, k)$ is given by

$$r(j,k)=[\pi(j)q(j,k)]^{-1}=[\pi(k)q(k,j)]^{-1}=r(k,j)$$

Consider now an electrical network with nodes $\mathcal{S}$ in which nodes j and k are connected by a wire of resistance $r(j, k)$ and node j is connected to earth by a capacitor of capacitance $\pi(j)$. If $u_j(t)$ is the charge present at node j at time t then $u_j(t)$, $j\in\mathcal{S}$, will satisfy equations (1.20); these are just Kirchhoff's equations and express the fact that the rate of increase of charge at node j is equal to the rate at which charge is flowing into node j. Thus the way in which probability spreads itself over the states of the Markov process is analogous to the way in which charge spreads itself over the nodes of the electrical network. Further, if we let $h(x)=-\frac{1}{2}x^2$ then

$$-H(t)=\frac{1}{2}\sum_{j\in\mathcal{S}}\frac{u_j(t)^2}{\pi(j)}$$

which is just the potential energy stored in the capacitors of the electrical network. As $H(t)$ increases, energy is dissipated as heat in the wires of the electrical network.

In this work we shall mainly be concerned with processes in equilibrium, exceptions being Section 4.5 and Chapter 5. In Chapter 5 the electrical analogue discussed here will be considered further.

Exercises 1.4

1. Show that the jump chain $X^J(t)$ of the Ehrenfest model has the same equilibrium distribution as $X(t)$. Show that if j is close to K, then in equilibrium

$$P(X^J(-1)=j-1, X^J(0)=j, X^J(1)=j-1)$$

is much larger than any of

$$P(X^J(-1)=j+1, X^J(0)=j, X^J(1)=j-1),$$

$$P(X^J(-1)=j-1, X^J(0)=j, X^J(1)=j+1),$$

$$P(X^J(-1)=j+1, X^J(0)=j, X^J(1)=j+1).$$

Deduce that if at a fixed time we observe j particles in container 1 then it is highly probable that the previous state was, and the next state will be, $j-1$.

2. If in the Ehrenfest model particles move from container 1 to container 2

at rate μ show that the equilibrium distribution is

$$\pi(j)=\left(1+\frac{\lambda}{\mu}\right)^{-K}\binom{K}{j}\left(\frac{\lambda}{\mu}\right)^{j}$$

3. Let $X(t)$ be a stationary stochastic process and let $\mathscr{A}$ be a subset of the state space $\mathscr{S}$. Show that

$$P(X(1), X(2), \ldots, X(n)\in \mathscr{A} \mid X(0)\in \mathscr{A}) = P(X(0), X(1), \ldots, X(n-1)\in \mathscr{A} \mid X(n)\in \mathscr{A})$$

Establish Kac's formula:

$$P(X(0)\in \mathscr{A}, X(1), X(2), \ldots, X(n)\notin \mathscr{A}) = P(X(0), X(1), \ldots, X(n-1)\notin \mathscr{A}, X(n)\in \mathscr{A})$$

Deduce that

$$P(X(1), X(2), \ldots, X(n)\notin \mathscr{A} \mid X(0)\in \mathscr{A}) = P(X(0), X(1), \ldots, X(n-1)\notin \mathscr{A} \mid X(n)\in \mathscr{A})$$

Observe that these relations hold whether the process is reversible or not.

4. Suppose the transition rates of a Markov process with a finite space satisfy the detailed balance conditions. If the process starts in state k at time $t=0$ show that

$$u_k(2t)=\sum_{j\in\mathscr{S}}\frac{\pi(k)}{\pi(j)}[u_j(t)]^2$$

Deduce from Theorem 1.6 that the function $u_k(t)$, $t\geq 0$, decreases monotonically from unity to $\pi(k)$.

1.5 KOLMOGOROV'S CRITERIA

The detailed balance conditions (1.6) enable us to decide whether a stationary Markov process is reversible from its equilibrium distribution and its transition rates. Since the equilibrium distribution is determined by the transition rates it is natural to ask whether we can establish the reversibility of a process directly from the transition rates alone. Kolmogorov's criteria allow us to do just that.

We begin by establishing the criteria for a Markov chain.

Theorem 1.7. *A stationary Markov chain is reversible if and only if its transition probabilities satisfy*

$$p(j_1, j_2)p(j_2, j_3)\cdots p(j_{n-1}, j_n)p(j_n, j_1) = p(j_1, j_n)p(j_n, j_{n-1})\cdots p(j_3, j_2)p(j_2, j_1) \qquad (1.21)$$

for any finite sequence of states $j_1, j_2, \ldots, j_n \in \mathscr{S}$.

Proof. If the process is reversible then the detailed balance conditions hold; hence

$$\begin{aligned}\pi(j_1)p(j_1, j_2) &= \pi(j_2)p(j_2, j_1)\\ \pi(j_2)p(j_2, j_3) &= \pi(j_3)p(j_3, j_2)\\ &\cdots\cdots\cdots\cdots\cdots\cdots\\ \pi(j_{n-1})p(j_{n-1}, j_n) &= \pi(j_n)p(j_n, j_{n-1})\\ \pi(j_n)p(j_n, j_1) &= \pi(j_1)p(j_1, j_n)\end{aligned}$$

Multiplying these conditions together and cancelling the positive equilibrium probabilities gives equation (1.21).

Conversely, suppose the transition probabilities satisfy equation (1.21). Let j_0 be an arbitrarily chosen reference state. Since the process is irreducible, for any state $j \in \mathscr{S}$ there exists a sequence of states $j, j_n, j_{n-1}, \ldots, j_1, j_0$ leading from j to j_0 such that $p(j, j_n)p(j_n, j_{n-1}) \cdots p(j_1, j_0) > 0$. Let

$$\pi(j) = B\frac{p(j_0, j_1)p(j_1, j_2) \cdots p(j_n, j)}{p(j, j_n)p(j_n, j_{n-1}) \cdots p(j_1, j_0)}$$

where B is a positive constant. Observe that $\pi(j)$ does not depend upon the particular sequence of states chosen to lead from j to j_0, since if j, $j'_m, j'_{m-1}, \ldots, j'_1, j_0$ is another sequence of states leading from j to j_0 with $p(j, j'_m)p(j'_m, j'_{m-1}) \cdots p(j'_1, j_0) > 0$, relation (1.21) ensures that

$$\frac{p(j_0, j_1)p(j_1, j_2) \cdots p(j_n, j)}{p(j, j_n)p(j_n, j_{n-1}) \cdots p(j_1, j_0)} = \frac{p(j_0, j'_1)p(j'_1, j'_2) \cdots p(j'_m, j)}{p(j, j'_m)p(j'_m, j'_{m-1}) \cdots p(j'_1, j_0)}$$

Note that irreducibility and relation (1.21) imply that $\pi(j)$ is positive. We must now show that $\pi(j)$, $j \in \mathscr{S}$, satisfy the detailed balance conditions. If $p(j, k) = p(k, j) = 0$ these are satisfied automatically, so suppose $p(k, j) > 0$. Then we can write

$$\pi(k) = B\frac{p(j_0, j_1)p(j_1, j_2) \cdots p(j_n, j)p(j, k)}{p(k, j)p(j, j_n)p(j_n, j_{n-1}) \cdots p(j_1, j_0)}$$

Hence

$$\pi(k)p(k, j) = \pi(j)p(j, k)$$

Thus $\pi(j)$, $j \in \mathscr{S}$, satisfy the detailed balance conditions and so they also satisfy the equilibrium equations. Since the process is stationary they cannot sum to infinity; hence B can be chosen so they sum to unity. Thus from Theorem 1.2 the process is reversible and $\pi(j)$, $j \in \mathscr{S}$, is the equilibrium distribution.

Kolmogorov's criteria (1.21) provide a useful insight into the nature of a reversible Markov chain. They show that given a starting point $j_1 \in \mathscr{S}$ any path in the state space which ultimately returns to j_1 must have the same probability whether this path is traced in one direction or the other. Thus a reversible Markov chain shows no net circulation in the state space.

The proof of Theorem 1.7 has a direct analogue for a Markov process which establishes the next result.

Theorem 1.8. *A stationary Markov process is reversible if and only if its transition rates satisfy*

$$q(j_1, j_2)q(j_2, j_3) \cdots q(j_{n-1}, j_n)q(j_n, j_1)$$
$$= q(j_1, j_n)q(j_n, j_{n-1}) \cdots q(j_3, j_2)q(j_2, j_1) \qquad (1.22)$$

for any finite sequence of states $j_1, j_2, \ldots, j_n \in \mathcal{S}$.

In practice relation (1.22) does not usually have to be established for all closed paths $j_1, j_2, \ldots, j_n, j_1$ since it is often possible to choose certain simple paths so that the truth of (1.22) for a general path follows from its truth for these simple paths. For instance if relation (1.22) can be established for sequences of distinct states then it follows for all sequences. Another example is contained in Exercise 1.5.2, and a further example follows.

A two-server queue. Suppose that the stream of customers arriving at a queue forms a Poisson process of rate ν and that there are two servers who possibly differ in efficiency. Specifically, suppose that a customer's service time at server i is exponentially distributed with mean μ_i^{-1}, for $i = 1, 2$, where to ensure that equilibrium is possible $\mu_1 + \mu_2 > \nu$. If a customer arrives to find both servers free he is equally likely to be allocated to either server. The queue can be represented by a Markov process whose transition rates and associated graph G are illustrated in Fig. 1.2. State n, for $n = 0, 2, 3, \ldots$, corresponds to there being n customers in the queue, while state $1A$ or $1B$ corresponds to there being a single customer in the queue, allocated to server 1 or 2 respectively. To ensure that the process is reversible in equilibrium we need only check the relation

$$q(0, 1A)q(1A, 2)q(2, 1B)q(1B, 0) = q(0, 1B)q(1B, 2)q(2, 1A)q(1A, 0) \qquad (1.23)$$

since Kolmogorov's criterion (1.22) for any other finite sequence of states

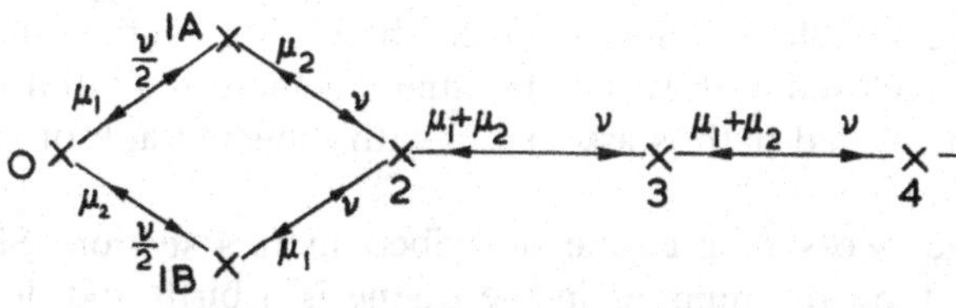

Fig. 1.2 Representation of a two-server queue

will follow from this or will hold trivially. Relation (1.23) holds, since it reduces to

$$\tfrac{1}{2}\nu \times \nu \times \mu_1 \times \mu_2 = \tfrac{1}{2}\nu \times \nu \times \mu_2 \times \mu_1$$

The equilibrium distribution is given by

$$\pi(1A) = \pi(0)\frac{\nu}{2\mu_1}$$

$$\pi(1B) = \pi(0)\frac{\nu}{2\mu_2}$$

$$\pi(n) = \pi(0)\frac{\nu^2}{2\mu_1\mu_2}\left(\frac{\nu}{\mu_1+\mu_2}\right)^{n-2} \qquad n = 2, 3, \ldots$$

Observe that if a customer arriving to find both servers free is allocated to server 1 with probability $p \neq \frac{1}{2}$ then the process is not reversible since relation (1.23) will fail to hold.

Exercises 1.5

1. There is an alternative proof of Theorem 1.7 which is instructive. By summing the equation

$$p(j, j_1)p(j_1, j_2) \cdots p(j_n, k)p(k, j) = p(j, k)p(k, j_n) \cdots p(j_2, j_1)p(j_1, j)$$

over all $j_1, j_2, \ldots, j_n \in \mathcal{S}$, and then letting $n \to \infty$, deduce that, for aperiodic chains in the first instance, the equilibrium distribution $\pi(j)$, $j \in \mathcal{S}$, satisfies

$$\pi(k)p(k, j) = \pi(j)p(j, k)$$

2. Consider a stationary Markov process with a state j_0 such that $q(j, j_0) > 0$ for all $j \in \mathcal{S}$. Show that a necessary and sufficient condition for reversibility is that

$$q(j_0, j_1)q(j_1, j_2)q(j_2, j_0) = q(j_0, j_2)q(j_2, j_1)q(j_1, j_0)$$

for all $j_1, j_2 \in \mathcal{S}$.

3. Construct a stationary Markov process which is not reversible yet which satisfies relation (1.22) when $n = 3$.
4. Consider a stationary Markov process whose associated graph G can be imbedded in the plane without any of its edges crossing. Show that the process is reversible if relation (1.22) holds for every minimal closed path, where a closed path is called minimal if there is a point in the plane such that the closed path is associated with the subgraph of G encircling the point.
5. Consider the two-server queue described in this section. Show that if $\mu_1 = \mu_2 = \mu$ then the number in the queue is a birth and death process and $\pi(0) = (2\mu - \nu)/(2\mu + \nu)$.

6. Generalize the queue described in this section to the case of s servers. Assume that if a customer arrives to find more than one server free he is equally likely to be allocated to any of them.
7. Observe that Lemma 1.5 could be regarded as a corollary of Theorem 1.8. Consider now the following amendment of the two-server queue described in this section. Suppose that if a customer arrives to find both servers free he is allocated to the server who has been free for the shortest time. Show that the resulting queue can be represented by a Markov process whose associated graph G is a tree. Generalize the queue to the case of s servers. Show that the probability servers $i_1, i_2, \ldots, i_m$ are busy and the rest free is the same as in the queue considered in the preceding exercise.

1.6 TRUNCATING REVERSIBLE PROCESSES

Various amendments can be made to the transition rates of a reversible Markov process without destroying the property of reversibility. For example if a reversible Markov process is altered by changing $q(j_1, j_2)$ to $cq(j_1, j_2)$ and $q(j_2, j_1)$ to $cq(j_2, j_1)$, where $c>0$, then the resulting Markov process is reversible and has the same equilibrium distribution. This follows from Theorem 1.3, since the detailed balance conditions (1.6) will still be satisfied. A slightly different alteration is the subject of the next lemma.

Lemma 1.9. *If the transition rates of a reversible Markov process with state space $\mathcal{S}$ and equilibrium distribution $\pi(j)$, $j \in \mathcal{S}$, are altered by changing $q(j, k)$ to $cq(j, k)$ for $j \in \mathcal{A}$, $k \in \mathcal{S} - \mathcal{A}$, where $c>0$, then the resulting Markov process is reversible in equilibrium and has equilibrium distribution*

$$\begin{aligned} &B\pi(j) && j \in \mathcal{A} \\ &Bc\pi(j) && j \in \mathcal{S} - \mathcal{A} \end{aligned}$$

where B is a normalizing constant.

Proof. The suggested equilibrium distribution satisfies the detailed balance conditions and so the result follows from Theorem 1.3. The normalizing constant is given by

$$B^{-1} = \sum_{j \in \mathcal{A}} \pi(j) + c \sum_{j \in \mathcal{S} - \mathcal{A}} \pi(j)$$

If $c=0$ the resulting process has a smaller state space. Say that a Markov process is *truncated* to the set $\mathcal{A} \subset \mathcal{S}$ if $q(j, k)$ is changed to zero for $j \in \mathcal{A}$, $k \in \mathcal{S} - \mathcal{A}$, and if the resulting process is irreducible within the state space $\mathcal{A}$. Like Lemma 1.9 the next result follows directly from the detailed balance conditions.

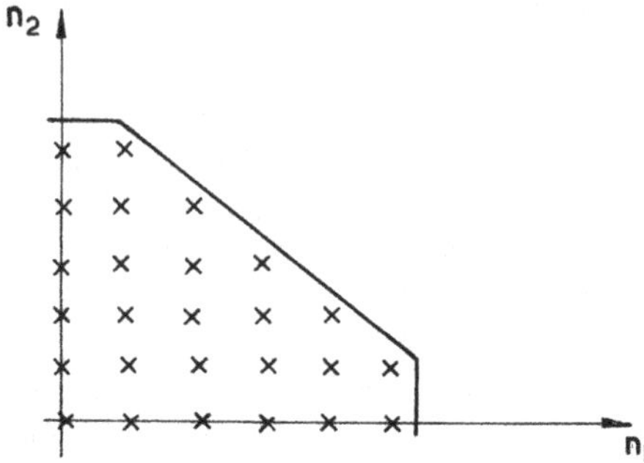

Fig. 1.3 The state space for two queues with a joint waiting room of size 4

Corollary 1.10. *If a reversible Markov process with state space $\mathcal{S}$ and equilibrium distribution $\pi(j)$, $j \in \mathcal{S}$, is truncated to the set $\mathcal{A} \subset \mathcal{S}$ then the resulting Markov process is reversible in equilibrium and has equilibrium distribution*

$$\frac{\pi(j)}{\sum_{k\in\mathcal{A}} \pi(k)} \qquad j \in \mathcal{A}$$

It is interesting to note that the equilibrium distribution of the truncated process is just the conditional probability that the original process is in state j given that it is somewhere in $\mathcal{A}$. An example has already been given in Exercise 1.3.2; another follows.

Two queues with a joint waiting room. Consider two independent $M/M/1$ queues. Let ν_i be the arrival rate and μ_i^{-1} the mean service time at queue i, for $i = 1, 2$. If n_i is the number of customers in queue i then the Markov process (n_1, n_2) is reversible (Exercise 1.2.8) with equilibrium distribution

$$\pi(n_1, n_2) = \left(1 - \frac{\nu_1}{\mu_1}\right)\left(\frac{\nu_1}{\mu_1}\right)^{n_1}\left(1 - \frac{\nu_2}{\mu_2}\right)\left(\frac{\nu_2}{\mu_2}\right)^{n_2}$$

Suppose now that the two queues are forced to share a joint waiting room of size R, so that a customer who arrives to find R customers already waiting for service, not including those being served, leaves without being served. This corresponds to truncating the Markov process (n_1, n_2) to $\mathcal{A}$, the set of states in which not more than R customers are waiting (Fig. 1.3). The equilibrium distribution for the truncated process will thus be

$$\pi(n_1, n_2) = \pi(0, 0)\left(\frac{\nu_1}{\mu_1}\right)^{n_1}\left(\frac{\nu_2}{\mu_2}\right)^{n_2} \qquad (n_1, n_2) \in \mathcal{A}$$

Exercises 1.6

1. Suppose that the two queues considered in this section have three waiting rooms associated with them: a waiting room of size R_1 for customers at

queue 1, a waiting room of size R_2 for customers at queue 2, and an overflow waiting room of size R_3 which can hold customers waiting for either queue. Identify the state space and write down the form of the equilibrium distribution.

2. Suppose that a Markov process with equilibrium distribution $\pi(j)$, $j \in \mathcal{S}$, is truncated to the set $\mathcal{A} \subset \mathcal{S}$. Show that the equilibrium distribution of the truncated process is the conditional probability distribution

$$\frac{\pi(j)}{\sum_{k \in \mathcal{A}} \pi(k)} \qquad j \in \mathcal{A}$$

if and only if the distribution $\pi(j)$, $j \in \mathcal{S}$, satisfies

$$\pi(j) \sum_{k \in \mathcal{A}} q(j, k) = \sum_{k \in \mathcal{A}} \pi(k) q(k, j) \qquad j \in \mathcal{A} \tag{1.24}$$

These equations are of a form intermediate between the detailed balance conditions (1.6) and the full balance conditions (1.3), and we shall call them the partial balance conditions for the set $\mathcal{A}$. Observe that the distribution $\pi(j)$, $j \in \mathcal{S}$, satisfies the partial balance conditions (1.24) if and only if

$$\pi(j) \sum_{k \in \mathcal{S} - \mathcal{A}} q(j, k) = \sum_{k \in \mathcal{S} - \mathcal{A}} \pi(k) q(k, j) \qquad j \in \mathcal{A}$$

These equations should be compared with equation (1.8).

3. Suppose that a Markov process with equilibrium distribution $\pi(j)$, $j \in \mathcal{S}$, is altered by changing the transition rate $q(j, k)$ to $cq(j, k)$ for $j, k \in \mathcal{A}$, where $c \neq 0$ or 1. Show that the resulting Markov process has the same equilibrium distribution if and only if the partial balance conditions (1.24) are satisfied.

4. Suppose that a Markov process with equilibrium distribution $\pi(j)$, $j \in \mathcal{S}$, is altered by changing the transition rate $q(j, k)$ to $cq(j, k)$ for $j \in \mathcal{A}$, $k \in \mathcal{S} - \mathcal{A}$, where $c \neq 0$ or 1. Show that the resulting Markov process has an equilibrium distribution of the form

$$\begin{array}{ll} B\pi(j) & j \in \mathcal{A} \\ Bc\pi(j) & j \in \mathcal{S} - \mathcal{A} \end{array}$$

if and only if the distribution $\pi(j)$, $j \in \mathcal{S}$, satisfies the partial balance conditions (1.24).

1.7 REVERSED PROCESSES

If $X(t)$ is a reversible Markov process then $X(\tau - t)$ is also a Markov process since it is statistically indistinguishable from $X(t)$. In this section we shall investigate the form of the reversed process $X(\tau - t)$ when $X(t)$ is a Markov process, but one which is not necessarily reversible.

The characterization of a Markov process as a process for which, conditional on the present, the past and the future are independent shows that if $X(t)$ is a Markov process then so is $X(\tau - t)$. An alternative proof is given in the next lemma which shows the complications that can arise if $X(t)$ is not stationary.

Lemma 1.11. *If $X(t)$ is a time homogeneous Markov process which is not stationary then the reversed process $X(\tau - t)$ is a Markov process which is not even time homogeneous.*

Proof. Since $X(t)$ is a Markov process we have the following factorization for $t_1 < t_2 < \cdots < t_n$:

$$P(j_1, j_2, \ldots, j_n) = P(j_1) \prod_{r=2}^{n} P(j_r \mid j_{r-1})$$

But

$$P(j_{r-1})P(j_r \mid j_{r-1}) = P(j_r)P(j_{r-1} \mid j_r) \tag{1.25}$$

and so

$$P(j_1, j_2, \ldots, j_n) = P(j_n) \prod_{r=2}^{n} P(j_{r-1} \mid j_r)$$

This factorization shows that $X(\tau - t)$ is Markov, but let us look more closely at the definition of $P(j_{r-1} \mid j_r)$ contained in equation (1.25). An alternative version of equation (1.25) is

$$\begin{aligned} &P(X(t) = j)P(X(t+h) = k \mid X(t) = j) \\ &\quad = P(X(t+h) = k)P(X(t) = j \mid X(t+h) = k) \end{aligned} \tag{1.26}$$

Now $P(X(t+h) = k \mid X(t) = j)$ does not depend upon t, but $P(X(t) = j)$ and $P(X(t+h) = k)$ will depend upon t for some $j, k \in \mathcal{S}$ if $X(t)$ is not stationary. Thus $P(X(t) = j \mid X(t+h) = k)$ will depend upon t, and so $X(\tau - t)$ will not be time homogeneous.

If $X(t)$ is stationary the situation is much simpler.

Theorem 1.12. *If $X(t)$ is a stationary Markov process with transition rates $q(j, k)$, $j, k \in \mathcal{S}$, and equilibrium distribution $\pi(j)$, $j \in \mathcal{S}$, then the reversed process $X(\tau - t)$ is a stationary Markov process with transition rates*

$$q'(j, k) = \frac{\pi(k)q(k, j)}{\pi(j)} \qquad j, k \in \mathcal{S}$$

and the same equilibrium distribution.

Proof. From equation (1.26) we obtain

$$P(X(t)=j \mid X(t+h)=k)=\frac{\pi(j)}{\pi(k)}P(X(t+h)=k \mid X(t)=j)$$

Now divide both sides by h and let h tend to zero. Thus

$$q'(k,j)=\frac{\pi(j)q(j,k)}{\pi(k)}$$

The fact that the reversed process is stationary follows as an immediate consequence of the definition of stationarity. That $X(t)$ and $X(\tau-t)$ have the same equilibrium distribution follows since they have the same stationary distribution, but it is worth checking that the equilibrium equations

$$\pi(j)\sum_{k\in\mathcal{S}} q'(j,k)=\sum_{k\in\mathcal{S}}\pi(k)q'(k,j)$$

are satisfied.

The next example illustrates the theorem.

A two-server queue. Suppose the stream of customers arriving at a two-server queue forms a Poisson process of rate ν and that a customer's service time at server i is exponentially distributed with mean μ_i^{-1}, for $i=1,2$, where $\mu_1+\mu_2>\nu$. If a customer arrives to find both servers free he is allocated to the server who has been free for the longest time. The queue can be represented by a Markov process whose transition rates and associated graph G are illustrated in Fig. 1.4(a). State n, for $n=2,3,\ldots,$

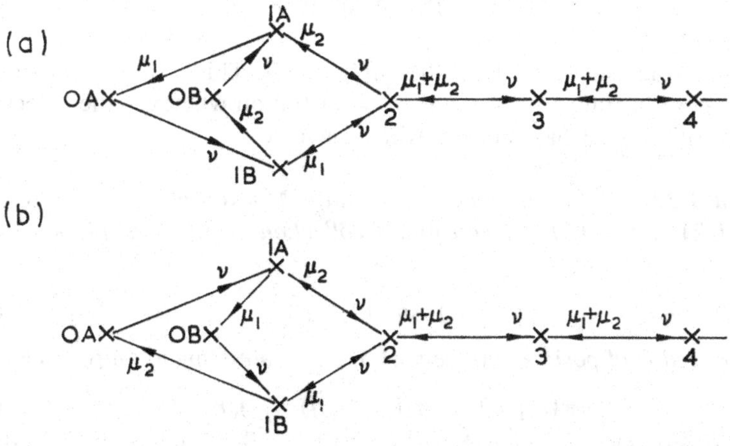

Fig. 1.4 A two-server queue: (a) the original process and (b) the reversed process

corresponds to there being n customers in the queue. State 1A or 1B corresponds to there being a single customer in the queue, allocated to server 1 or 2 respectively. State 0A or 0B corresponds to both servers being free, with server 1 or 2 respectively having been free for the shorter time. The process is clearly not reversible since $q(0\text{A}, 1\text{B})$ is positive and $q(1\text{B}, 0\text{A})$ is zero. The equilibrium distribution for the process is

$$\pi(n) = \pi(2)\left(\frac{\nu}{\mu_1+\mu_2}\right)^{n-2} \qquad n = 2, 3, \ldots$$

$$\pi(1\text{A}) = \pi(2)\frac{\mu_2}{\nu}$$

$$\pi(1\text{B}) = \pi(2)\frac{\mu_1}{\nu}$$

$$\pi(0\text{A}) = \pi(0\text{B}) = \pi(2)\frac{\mu_1\mu_2}{\nu^2}$$

Theorem 1.12 shows that the transition rates of the reversed process are as illustrated in Fig. 1.4(b). Observe that they take a particularly simple form. This is not always the case, as Exercise 1.7.1 demonstrates.

Remember that the period for which $X(t)$ remains in state j is exponentially distributed with parameter

$$q(j) = \sum_{k\in\mathcal{S}} q(j, k)$$

Similarly, define

$$q'(j) = \sum_{k\in\mathcal{S}} q'(j, k)$$

It follows from Theorem 1.12 that $q(j) = q'(j)$. This is not surprising: the periods spent in state j have the same distribution whatever the direction of time. Theorem 1.12 has the following converse.

Theorem 1.13. *Let $X(t)$ be a stationary Markov process with transition rates $q(j, k)$, $j, k \in \mathcal{S}$. If we can find a collection of numbers $q'(j, k)$, $j, k \in \mathcal{S}$, such that*

$$q'(j) = q(j) \qquad j \in \mathcal{S} \tag{1.27}$$

and a collection of positive numbers $\pi(j)$, $j \in \mathcal{S}$, summing to unity, such that

$$\pi(j)q(j, k) = \pi(k)q'(k, j) \qquad j, k \in \mathcal{S} \tag{1.28}$$

then $q'(j, k)$, $j, k \in \mathcal{S}$, are the transition rates of the reversed process $X(\tau - t)$ and $\pi(j)$, $j \in \mathcal{S}$, is the equilibrium distribution of both processes.

Proof. From equations (1.28) and (1.27) it follows that

$$\sum_{j\in\mathcal{S}} \pi(j)q(j,k) = \pi(k)\sum_{j\in\mathcal{S}} q'(k,j)$$
$$= \pi(k)q'(k)$$
$$= \pi(k)q(k)$$

Thus $\pi(j)$, $j \in \mathcal{S}$, is the equilibrium distribution of $X(t)$. That $q'(j,k)$, $j, k \in \mathcal{S}$, are the transition rates of the reversed process then follows from Theorem 1.12.

We shall find Theorem 1.13 useful in Chapter 3 where we discuss a rather complicated Markov process for which it would be tedious to check the equilibrium equations, but for which possible transition rates of the reversed process are apparent. The similarity of equation (1.28) to the detailed balance condition should be observed. A generalization of Kolmogorov's criteria can also be obtained (Exercise 1.7.4).

Occasionally we may come across a stationary Markov process for which the reversed process, while not statistically indistinguishable from the original process, would be if some of the states were interchanged. To make this notion precise suppose that to each state $j \in \mathcal{S}$ there corresponds a conjugate state $j^+ \in \mathcal{S}$ with $(j^+)^+ = j$. Then the stationary Markov process $X(t)$ is called *dynamically reversible* if $X(t)$ is statistically indistinguishable from $[X(\tau - t)]^+$. As an example consider the stationary Markov process with state space $\mathcal{S} = \{-n, -n+1, \ldots, n-1, n\}$ and transition rates

$$q(j, j+1) = \lambda \qquad j = -n, -n+1, \ldots, n-1$$
$$q(n, -n) = \lambda$$

With $j^+ = -j$ this process is dynamically reversible. Reversing this process has an analogous effect to reversing the velocity of a particle moving in a circular orbit—hence the term 'dynamically reversible'. A further example is the two-server queue illustrated in Fig. 1.4, which is dynamically reversible with $(0A)^+ = 0B$ and all other states self-conjugate.

Theorem 1.14. *A stationary Markov process with* $q(j) = q(j^+)$, $j \in \mathcal{S}$, *is dynamically reversible if and only if there exists a collection of positive numbers* $\pi(j)$, $j \in \mathcal{S}$, *summing to unity that satisfy*

$$\pi(j) = \pi(j^+) \qquad j \in \mathcal{S} \tag{1.29}$$

and

$$\pi(j)q(j,k) = \pi(k^+)q(k^+, j^+) \qquad j, k \in \mathcal{S}$$

When there exists such a collection $\pi(j)$, $j \in \mathcal{S}$, *it is the equilibrium distribution of the process.*

Proof. If the process is dynamically reversible then the relations follow from the identification $\pi(j)=P(X(t)=j)$. Conversely, if the process satisfies the relations let $q'(k,j)=q(k^+,j^+)$. Thus

$$q'(k,j)=\frac{\pi(j)}{\pi(k^+)}q(j,k)$$
$$=\frac{\pi(j)}{\pi(k)}q(j,k)$$

Further,

$$q'(j)=\sum_{k\in\mathcal{S}}q'(j,k)$$
$$=\sum_{k\in\mathcal{S}}q(j^+,k^+)$$
$$=q(j^+)$$
$$=q(j)$$

We have thus established that the transition rates $q'(j,k)$, $j,k\in\mathcal{S}$, satisfy equations (1.27) and (1.28) and so, by Theorem 1.13, $\pi(j)$, $j\in\mathcal{S}$, is the equilibrium distribution and the reversed process $X(\tau-t)$ has transition rates $q'(j,k)$, $j,k\in\mathcal{S}$. Since $q'(j,k)=q(j^+,k^+)$ the process $X(t)$ is dynamically reversible.

Exercises 1.7

1. If $X(t)$ is the stationary Markov process whose transition rates were given in Exercise 1.1.4, with $a\leq 1$ and $b>0$, find the transition rates of the reversed process $X(\tau-t)$.
2. Construct examples to show that condition (1.27) cannot be dropped from Theorem 1.13, nor condition (1.29) from Theorem 1.14.
3. Establish counterparts of Theorems 1.12, 1.13, and 1.14 for Markov chains. Observe that no analogue of condition (1.27) is needed: the implicit condition that transition probabilities sum to unity serves the same purpose.
4. Let $X(t)$ be a stationary Markov chain with transition probabilities $p(j,k)$, $j,k\in\mathcal{S}$. Show that if there exist transition probabilities $p'(j,k)$, $j,k\in\mathcal{S}$, such that

$$p(j_1,j_2)p(j_2,j_3)\cdots p(j_{n-1},j_n)p(j_n,j_1)$$
$$=p'(j_1,j_n)p'(j_n,j_{n-1})\cdots p'(j_3,j_2)p'(j_2,j_1)$$

for any finite sequence of states $j_1,j_2,\ldots,j_n\in\mathcal{S}$, then $p'(j,k)$, $j,k\in\mathcal{S}$, are the transition probabilities of the reversed Markov chain $X(\tau-t)$. Using the additional condition (1.27) obtain the parallel result for a Markov process.

5. Show that the reversed process illustrated in Fig. 1.4(b) can be regarded as representing a two-server queue identical to the one represented by the original process but with states 0A or 0B indicating that the next arrival will be allocated to server 1 or 2 respectively.
6. Generalize the queue considered in this section to the case of s servers. Show that the probability servers $i_1, i_2, \ldots, i_m$ are busy and the rest free is the same as in the queues considered in Exercises 1.5.6 and 1.5.7.
7. Suppose that a Markov process $X(t)$ with transition rates $q(j, k)$, $j, k \in \mathcal{S}$, and equilibrium distribution $\pi(j)$, $j \in \mathcal{S}$, is truncated to the set $\mathcal{A}$. Let $Y(t)$ be the stationary truncated process. Let $Z(t)$ be the stationary process resulting from truncating the reversed process $X(-t)$ to the set $\mathcal{A}$. Show that $Z(t)$ and $Y(-t)$ have the same transition rates if and only if the partial balance conditions (1.24) are satisfied. If $Z(t)$ and $Y(-t)$ have the same transition rates we shall say that for the process $X(t)$ the operations of time reversal and truncation to the set $\mathcal{A}$ commute.
8. Consider a Markov process with transition rates $q(j, k)$, $j, k \in \mathcal{S}$, and equilibrium distribution $\pi(j)$, $j \in \mathcal{S}$. Suppose that the probability flux out of the set $\mathcal{A}$

$$\sum_{j \in \mathcal{A}} \sum_{k \in \mathcal{S} - \mathcal{A}} \pi(j) q(j, k)$$

is finite. Show that the Markov chain formed by observing the process at those instants in time just before it leaves the set $\mathcal{A}$ has the same equilibrium distribution as the Markov chain formed by observing the process at those instants in time just after it enters the set $\mathcal{A}$ if and only if the partial balance conditions (1.24) are satisfied.

CHAPTER 2

Migration Processes

In this chapter we shall meet some of the simpler systems in which customers (or individuals) move about between a number of queues (or colonies). First we shall consider further the simple queue introduced in Section 1.3.

2.1 THE OUTPUT FROM A SIMPLE QUEUE

In Section 1.3 it was shown that if $n(t)$ is the number of customers in an $M/M/1$ queue at time t then in equilibrium $n(t)$ is a reversible Markov process. A typical realization of $n(t)$ is illustrated in Fig. 2.1. Note that the points in time at which $n(t)$ jumps upwards form a Poisson process of rate ν since these points correspond to arrivals at the queue. Now $n(t)$ is reversible and hence the points in time at which $n(-t)$ jumps upwards must also form a Poisson process of rate ν. But if $n(-t)$ jumps upwards at time $-t_0$ then $n(t)$ jumps downwards at time t_0, and so the points in time at which $n(t)$ jumps downwards must form a Poisson process of rate ν. But these points correspond to departures from the queue. We have thus shown that in equilibrium the points in time at which customers leave the queue (the departure process) form a Poisson process of rate ν. The line of argument can be used to establish a little more. Let t_0 be a fixed instant in time. Since $n(t)$ is reversible, the departure process up until time t_0 and the number in the queue at time t_0 have the same joint distribution as the arrival process after time $-t_0$ and the number in the queue at time $-t_0$. But the arrival process after time $-t_0$ is independent of the number in the queue at time $-t_0$, and hence the departure process prior to time t_0 is independent of the number in the queue at time t_0. The next theorem summarizes these results.

Theorem 2.1. *In equilibrium the departure process from an $M/M/1$ queue is a Poisson process, and the number in the queue at time t_0 is independent of the departure process prior to time t_0.*

In some ways this result is surprising, since while the server is busy departures occur at rate μ and while the server is idle departures occur at rate zero. It is difficult, however, to analyse the departure process using this approach since the length of a busy period and the departure process during this period are not independent. The dependence is such that if we observe

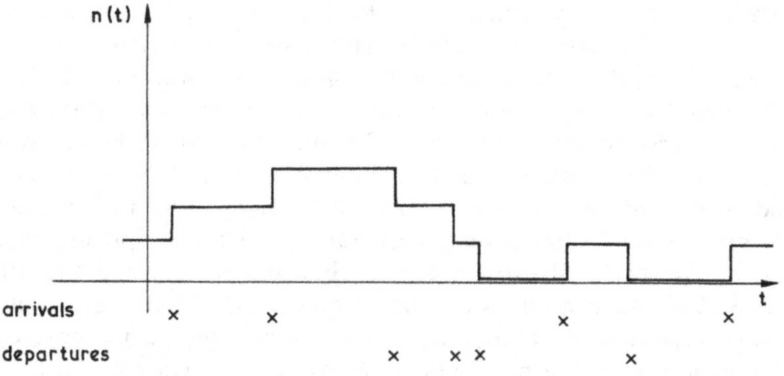

Fig. 2.1 A realization of the process $n(t)$

the entire departure process from an $M/M/1$ queue for $-\infty<t<\infty$, but know nothing of the times of arrival or the numbers in the queue, then we can determine the arrival rate ν but can learn nothing of the service rate μ.

The reasoning which led to Theorem 2.1 will apply to any queue with a Poisson arrival process for which the number in the queue is a birth and death process, for example the $M/M/s$ queue (Exercise 1.3.3). More generally it will apply whenever a queue with a Poisson arrival process can be represented by a reversible Markov process, provided an arrival causes the process to change state and the reverse transition corresponds to a departure. A further example of such a queue is the two-server queue discussed in Section 1.5. It occasionally requires some guile to find an appropriate process, as the following example illustrates.

A telephone exchange. Consider the model of a telephone exchange with K lines described in Section 1.3. The number of calls in progress at time t, n, is a reversible Markov process, but one which does not always change state when a call is initiated. Consider, however, the process (n, f) where the flip-flop variable f takes the value zero or unity and changes value whenever a call is lost. Clearly this process changes state whenever a call is initiated, and it is easily checked that the process is reversible with equilibrium distribution

$$\pi(n, f)=\tfrac{1}{2}\pi(n) \qquad n=0, 1, \ldots, K; \quad f=0, 1$$

where $\pi(n)$ is the equilibrium distribution of the process n. Moreover, transitions of the process associated with the completion of a call or the loss of a call are just the reverse transitions of those associated with the initiation of a call. Thus the points in time at which a call is lost or is completed form a Poisson process. If the points in time at which a call is lost are considered alone they form a more complicated point process, but one which is reversible (Exercise 2.1.3).

A two-server queue. Consider now the two-server queue introduced in Section 1.7. The Markov process representing this queue (Fig. 1.4a) is not reversible. Nevertheless, we do know the form of the reversed process (Fig. 1.4b). Indeed the reversed process can be regarded as representing an identical two-server queue but with a different interpretation being given to states $0A$ and $0B$ (Exercise 1.7.5). Observe that if a transition in the reversed process corresponds to an arrival then the reverse transition in the original process corresponds to a departure. Arrivals at the queue represented by the reversed process form a Poisson process and the arrival process at this queue after time $-t_0$ is independent of the state of the reversed process at time $-t_0$. Hence departures from the queue represented by the original process form a Poisson process and the state of the original process at time t_0 is independent of the departure process prior to time t_0. This example shows that it is not reversibility as such that leads to the results, but rather the particular form of the reversed process.

Exercises 2.1

1. Consider a queue with s identical servers who each take an exponentially distributed amount of time to serve a customer. Suppose that an arriving customer leaves immediately without being served (he *balks*), with a probability depending on the number in the queue, and that if he does join the queue he gives up and defects after an exponentially distributed amount of time unless his service has begun beforehand. Use both of the following approaches to show that if the arrival process is Poisson then in equilibrium the departure process is Poisson, provided all departing customers are counted.
   (i) Represent the queue by a Markov process (n, f) as in the telephone exchange model.
   (ii) Approximate the queue by one at which customers who decide on arrival that they will leave without service remain in the queue for an exponentially distributed time with mean ξ^{-1} where ξ is very large. Let m be the number of such customers in the queue. Suppose that while m is positive service and defection are suspended and further arrivals decide to leave the queue without service, i.e. they increase m. Let n be the number of other customers in the queue. Find the equilibrium distribution of the Markov process (n, m).
2. Show that the departure process from the queue considered in the preceding exercise remains Poisson if the defection rate of a customer depends upon how many are in front of him in the queue. Show that the departure processes from the many-server queues considered in Exercises 1.5.6, 1.5.7, and 1.7.6 are Poisson and remain so even if customers may balk or defect.

3. Before we can assert that a point process is reversible we need to characterize a point process in a form to which our definition of reversibility (Section 1.2) can apply. The simplest way to do this is with a flip-flop variable $f(t)$ defined as follows: $f(t)$ takes the value zero or unity and changes value at the points in time of the point process. Call the point process reversible if $f(t)$ is reversible. Show that in the model of a telephone exchange the intervals between successive lost calls are independent, and deduce that the points in time at which a call is lost form a reversible point process. (A point process in which the intervals between successive points are independent is called a renewal process.) Observe that a stationary renewal process is always reversible.
4. In the model of a telephone exchange show that the points in time at which a call is completed form a point process which when reversed in time is statistically indistinguishable from that formed by the points in time at which a call is successfully connected. Show that without the time reversal the two processes will differ unless $K=1$.

2.2 A SERIES OF SIMPLE QUEUES

The most obvious application of Theorem 2.1 is to a series of J single-server queues arranged so that when a customer leaves a queue he joins the next one, until he has passed through all queues (Fig. 2.2). Suppose the arrival stream at queue 1 is Poisson at rate ν and that service times at queue j are exponentially distributed with mean μ_j^{-1}, where $\nu<\mu_j$ for $j=1,2,\ldots,J$. Suppose further that service times are independent of each other, including those of the same customer in different queues, and of the arrival stream at queue 1. Let $n_j(t)$ be the number of customers in queue j at time t. Queue 1 viewed in isolation is simply an $M/M/1$ queue and hence the departure process from it is Poisson, by Theorem 2.1. Thus the arrival process at queue 2 is Poisson, and so it, too, viewed in isolation, is an $M/M/1$ queue. Proceeding with this argument we see that queue j viewed in isolation is an $M/M/1$ queue, and hence in equilibrium

$$\pi_j(n_j)=\left(1-\frac{\nu}{\mu_j}\right)\left(\frac{\nu}{\mu_j}\right)^{n_j}$$

What is not yet clear is the joint distribution of $(n_1, n_2, \ldots, n_J)$. Now Theorem 2.1 also states that $n_1(t_0)$ is independent of the departure process from queue 1 prior to t_0. But $(n_2(t_0), n_3(t_0), \ldots, n_J(t_0))$ is determined by the

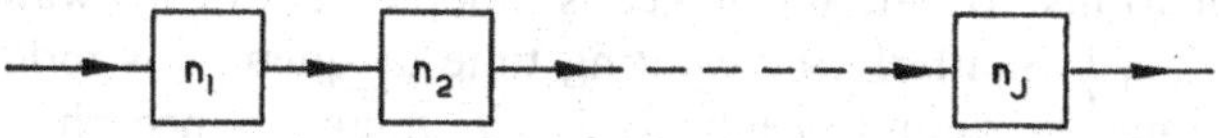

Fig. 2.2 A series of queues

departure process from queue 1 prior to t_0 and service times at queues $2, 3, \ldots, J$. Hence $n_1(t_0)$ is independent of $(n_2(t_0), n_3(t_0), \ldots, n_J(t_0))$. Similarly, $n_j(t_0)$ is independent of $(n_{j+1}(t_0), \ldots, n_J(t_0))$. Thus $n_1(t_0)$, $n_2(t_0), \ldots,$ $n_J(t_0)$ are mutually independent, and so in equilibrium

$$\pi(n_1, n_2, \ldots, n_J) = \prod_{j=1}^{J} \left(1 - \frac{\nu}{\mu_j}\right)\left(\frac{\nu}{\mu_j}\right)^{n_j}$$

The above approach is clearly of much wider applicability. The queues in the system can be of any of the forms discussed in the last section, and indeed the final queue need not be restricted even in this way. It is not essential that customers who leave queue j should join queue $j+1$; they may leave the system or jump to a queue between $j+1$ and J. We shall not pursue this approach, however, since it breaks down when a customer leaving queue j is allowed to jump back to a queue between 1 and j. Such behaviour will be discussed in the following sections.

Consider now the experience of an individual customer as he passes through the series of J simple queues described at the beginning of this section.

Theorem 2.2. *If the discipline at each queue in a series of J simple queues is first come first served, then in equilibrium the waiting times of a customer at each of the J queues are independent.*

Proof. The first step of the proof is to establish that in equilibrium the waiting time of a customer at a first come first served $M/M/1$ queue is independent of the departure process prior to his departure. Let $n(t)$ be the number of customers in the queue at time t. Then $n(-t)$ can also be regarded as the number in a first come first served $M/M/1$ queue at time t, since its behaviour is statistically indistinguishable from that of $n(t)$. Now if a customer arrives at the original queue at time t_0 and leaves at time t_1 then $n(-t)$ will signal the arrival of a customer at time $-t_1$ and the departure of the *same* customer at time $-t_0$. But the waiting time of this customer is independent of the arrivals signalled by $n(-t)$ after time $-t_1$. Hence in the original queue the departure process prior to time t_1 is independent of the waiting time of the customer who leaves at time t_1.

Consider a customer leaving queue 1. Customers who leave queue 1 after him cannot reach any subsequent queue before him: the queue discipline and the assumption of a single server at the next $J-2$ queues ensure this. Now his waiting time at queue 1 is independent of the arrival process at queue 2 prior to his arrival, and hence is independent of his waiting time at queues $2, 3, \ldots, J$. Similarly, his waiting time at queue j is independent of his waiting times at queues $j+1$, $j+2, \ldots, J$, and hence the theorem is proved.

It is clear from the proof of Theorem 2.2 that the final queue in the system is not required to be simple. For example waiting times would still be independent if the Jth queue were a first come first served $M/G/s$ queue, i.e. an s-server queue at which service times have a general distribution. Few other generalizations are possible; the independence of waiting times is a much less common result than the independence of queue sizes.

Exercises 2.2

1. If in a series of simple queues $\mu_1 = \mu_2 = \cdots = \mu_J$ show that the Markov process $(n_1, n_2, \ldots, n_J)$ is dynamically reversible.
2. Observe that in a series of simple queues the waiting time of a customer at queue j is exponential with mean $(\mu_j - \nu)^{-1}$. Deduce that the time taken for a customer to pass through the system is the sum of J independent exponentially distributed random variables, and has mean $\sum_j (\mu_j - \nu)^{-1}$ and variance $\sum_j (\mu_j - \nu)^{-2}$.
3. Consider two stacks, as described in Exercise 1.3.8, arranged so that customers leaving the first stack join the second. Show that in equilibrium the waiting time of a customer at the first stack is independent of the departure process *subsequent* to his departure. Deduce that the waiting times of a customer at the two stacks are independent.
4. Let $n(t)$ be the number of customers in an $M/M/s$ queue at time t. Suppose the queue discipline is first come first served, and let t_0 and t_1 be points in time at which $n(t)$ increases and decreases respectively. From the realization $n(t)$, $-\infty < t < \infty$, the probability P that the customer arriving at time t_0 is the one leaving at time t_1 can be calculated. Note that P will be zero or unity if $s = 1$. If the reversed process $n(-t)$ is regarded as representing the number in a first come first served $M/M/s$ queue, show that P is the probability that in this queue the customer who arrives at time $-t_1$ is the one who leaves at time $-t_0$. Deduce that in equilibrium the waiting time of a customer at a first come first served $M/M/s$ queue is independent of the departure process prior to his departure.
5. Consider a series of J first come first served $M/M/s$ queues in equilibrium. Let s_j be the number of servers at queue j. Deduce from the previous exercise that the waiting times of a customer at two successive queues are independent. Consider the case $J \geq 3$, $s_1 = s_3 = 1$, $s_2 = \infty$, $\mu_1 = \mu_2 = \mu_3$. Show that if a customer's waiting time at queue 1 is large then the probability that the customer entering queue 1 after him will overtake him and be present in queue 3 when he arrives there is close to one-eighth. Deduce that although a customer's waiting times at queues 1 and 2 or at queues 2 and 3 are independent, his waiting times at queues 1 and 3 are dependent. Deduce from the previous exercise that if $s_j = 1$

unless $j=1$ or J then the waiting times of a customer at each of the J queues are independent.

6. Consider a series of two simple queues in equilibrium. Suppose that an arriving customer finds queue 1 empty. Show that the probability queue 2 will be empty when he reaches it is

$$1-\frac{\nu}{\mu_2}+\frac{\nu}{\mu_2}\left(\frac{\mu_2-\nu}{\mu_1+\mu_2-\nu}\right)$$

Deduce that although a customer's waiting times at the two queues are independent his queueing times are not.

2.3 CLOSED MIGRATION PROCESSES

The elegant but delicate method of analysis used in the preceding sections breaks down if customers can rejoin queue 1 after leaving queue J. In this and the next section we shall use an alternative approach which can deal with such behaviour. The approach readily yields equilibrium distributions for the number of customers in each queue, but is not as informative about the time taken by a customer to pass through a sequence of queues.

We shall call the model to be examined a migration process. The main applications are to queueing rather than to biological systems, but the idea of individuals moving between colonies makes exposition easier and the alternative term 'queueing network' seems more appropriate for the model of the next chapter. In this section we shall consider a closed migration process where individuals cannot enter or leave the system but can only move between colonies. Thus the total number of individuals in the system, N, is fixed.

Consider a set of J colonies and let n_j denote the number of individuals in colony j, for $j=1, 2, \ldots, J$. Define an operator T_{jk} to act upon the vector $\mathbf{n}=(n_1, n_2, \ldots, n_J)$ as follows:

$$T_{jk}(n_1, n_2, \ldots, n_j, \ldots, n_k, \ldots, n_J)=(n_1, n_2, \ldots, n_j-1, \ldots, n_k+1, \ldots, n_J)$$

if $j<k$. Similarly,

$$T_{jk}(n_1, n_2, \ldots, n_k, \ldots, n_j, \ldots, n_J)=(n_1, n_2, \ldots, n_k+1, \ldots, n_j-1, \ldots, n_J)$$

if $k<j$. Thus T_{jk} moves an individual from colony j to k. We shall study $\mathbf{n}$ under the assumption that it is a Markov process with transition rates given by

$$q(\mathbf{n}, T_{jk}\mathbf{n})=\lambda_{jk}\phi_j(n_j) \tag{2.1}$$

where $\phi_j(0)=0$ and for simplicity $\lambda_{jj}=0$. The parameter λ_{jk} can be viewed as measuring the intrinsic tendency for movement from colony j to colony k; the function $\phi_j(n)$ then measures the extent to which this tendency is affected by the number of individuals in colony j. To ensure that $\mathbf{n}$ is irreducible within the state space

$$\mathcal{S}=\left\{\mathbf{n} \mid n_j \geq 0, j=1,2,\ldots,J; \sum_{j=1}^{J} n_j = N\right\}$$

we shall require that $\phi_j(n)>0$ if $n>0$ and that the parameters λ_{jk} allow an individual to pass between any two colonies, either directly or indirectly via a chain of other colonies. We shall call the process $\mathbf{n}$ a closed migration process.

As an example of the behaviour transition rates (2.1) can allow consider the special case

$$\phi_j(n)=\min(n, s)$$

With this function colony j behaves as a queue with s servers at which service times are exponentially distributed with mean λ_j^{-1}, where

$$\lambda_j=\sum_k \lambda_{jk}$$

An individual departing from this queue joins colony k with probability λ_{jk}/λ_j.

If $\phi_j(n)=n$ for all j then the migration process is called linear, and the individuals can be considered to be moving independently of one another. If $N=1$ then the single individual in the system performs a random walk on the set of colonies, and if $\alpha_1, \alpha_2, \ldots, \alpha_J$ is the unique collection of positive numbers summing to unity which satisfy

$$\alpha_j \sum_k \lambda_{jk}=\sum_k \alpha_k \lambda_{kj} \qquad j=1,2,\ldots,J \tag{2.2}$$

then α_j is the equilibrium probability that the individual is in colony j.

Theorem 2.3. *The equilibrium distribution for a closed migration process is*

$$\pi(\mathbf{n})=B_N \prod_{j=1}^{J} \frac{\alpha_j^{n_j}}{\prod_{r=1}^{n_j} \phi_j(r)} \qquad \mathbf{n}\in\mathcal{S} \tag{2.3}$$

where B_N is a normalizing constant, chosen so that the distribution sums to unity.

Proof. The equilibrium equations (1.3) are

$$\pi(\mathbf{n}) \sum_{j=1}^{J} \sum_{k=1}^{J} q(\mathbf{n}, T_{jk}\mathbf{n})=\sum_{j=1}^{J} \sum_{k=1}^{J} \pi(T_{jk}\mathbf{n})q(T_{jk}\mathbf{n}, \mathbf{n})$$

which become

$$\pi(\mathbf{n}) \sum_{j=1}^{J} \sum_{k=1}^{J} \lambda_{jk}\phi_j(n_j) = \sum_{j=1}^{J} \sum_{k=1}^{J} \pi(T_{jk}\mathbf{n})\lambda_{kj}\phi_k(n_k+1) \tag{2.4}$$

These will be satisfied if we can find a distribution $\pi(\mathbf{n})$, $\mathbf{n} \in \mathcal{S}$, which satisfies

$$\pi(\mathbf{n}) \sum_{k=1}^{J} \lambda_{jk}\phi_j(n_j) = \sum_{k=1}^{J} \pi(T_{jk}\mathbf{n})\lambda_{kj}\phi_k(n_k+1) \tag{2.5}$$

If $n_j = 0$ then, with the convention that $\pi(\mathbf{n})$ vanishes if $\mathbf{n} \notin \mathcal{S}$, equations (2.5) are satisfied trivially. When $n_j > 0$ it is readily verified, using equation (2.2), that the form proposed for $\pi(\mathbf{n})$ satisfies equations (2.5). Thus $\pi(\mathbf{n})$, $\mathbf{n} \in \mathcal{S}$, satisfies the equilibrium equations (2.4) and, since $\mathcal{S}$ is finite, it is clearly possible to choose B_N so that the distribution sums to unity.

The process $\mathbf{n}$ will be reversible if $\alpha_1, \alpha_2, \ldots, \alpha_J$ satisfy

$$\alpha_j\lambda_{jk} = \alpha_k\lambda_{kj}$$

since then the detailed balance conditions

$$\pi(\mathbf{n})\lambda_{jk}\phi_j(n_j) = \pi(T_{jk}\mathbf{n})\lambda_{kj}\phi_k(n_k+1) \tag{2.6}$$

will hold. The relations (2.5) are of a form intermediate between the detailed balance conditions (2.6) and the full balance conditions (2.4). We shall call them partial balance equations. Their connection with the partial balance conditions defined in Exercise 1.6.2 will be explored in Chapter 9; in this chapter our only use of partial balance will be to simplify the verification of equilibrium distributions.

The partial balance equations (2.5) state that in equilibrium the probability flux out of a state due to an individual moving from colony j is equal to the probability flux into that same state due to an individual moving to colony j. This statement is *not* clear a priori, and should be contrasted with the balance equation

$$\sum_{\mathbf{n}\in\mathcal{S}} \pi(\mathbf{n}) \sum_{k=1}^{J} q(\mathbf{n}, T_{jk}\mathbf{n}) = \sum_{\mathbf{n}\in\mathcal{S}} \sum_{k=1}^{J} \pi(T_{jk}\mathbf{n})q(T_{jk}\mathbf{n}, \mathbf{n}) \tag{2.7}$$

which states that in equilibrium the probability flux that an individual leaves colony j is equal to the probability flux that an individual enters colony j. This statement *is* clear a priori (and holds even if the transition rates (2.1) take a more general form) since in equilibrium the mean arrival rate at colony j must equal the mean departure rate from colony j.

Note that if λ_{jk}, $k = 1, 2, \ldots, J$, are decreased by a constant factor and $\phi_j(n)$, $n = 0, 1, 2, \ldots$, is increased by the same factor, then neither the transition rates (2.1) nor the equilibrium distribution $\pi(\mathbf{n})$ are altered in value. Note also that if the solution $\alpha_1, \alpha_2, \ldots, \alpha_J$ of equations (2.2) is not normalized to sum to unity the expression (2.3) remains valid; the normalizing constant B_N will alter accordingly. These observations can often simplify manipulations, but the task of determining B_N usually remains computationally tedious.

An important class of closed migration processes have the following property:

$$\lambda_{jk} = 0 \qquad \text{unless } k = j+1$$

and

$$\lambda_{j,j+1} = 1$$

for $j = 1, 2, \ldots, J-1$

$$\lambda_{Jk} = 0 \qquad \text{unless } k = 1$$

and

$$\lambda_{J1} = 1$$

Thus an individual repeatedly moves around the cycle of colonies $1, 2, \ldots, J, 1, 2, \ldots$. Such processes are called cyclic queues, and we shall devote the rest of this section to some examples of them.

The provision of spare components. Suppose that there are s_1 machines which each require a certain component in order to operate. A component in use fails after a period which is exponentially distributed with mean ϕ_1^{-1}. It is then replaced from a store of available components unless this is empty, in which case the machine lies idle until a component becomes available. There are s_2 servicing facilities to deal with failed components, and the length of time taken to service a component is exponentially distributed with mean ϕ_2^{-1}. After being serviced a component is returned to the store of available components. There are a total of N components altogether, and an issue of interest is the extent to which increasing N reduces the idle time of the machines.

If we regard the components as customers the system is equivalent to a cyclic queue with

$$\phi_j(n_j) = \phi_j \min(n_j, s_j) \qquad j = 1, 2$$

where n_1 is the number of components in use and in store, and $n_2 = N - n_1$. For a cyclic queue a solution of equation (2.2) is $\alpha_1 = \alpha_2 = \cdots = \alpha_J = 1$ and so Theorem 2.3 shows that the equilibrium distribution is

$$\pi(n_1, n_2) = \frac{B_N}{\prod_{r=1}^{n_1} \phi_1(r) \prod_{r=1}^{n_2} \phi_2(r)}$$

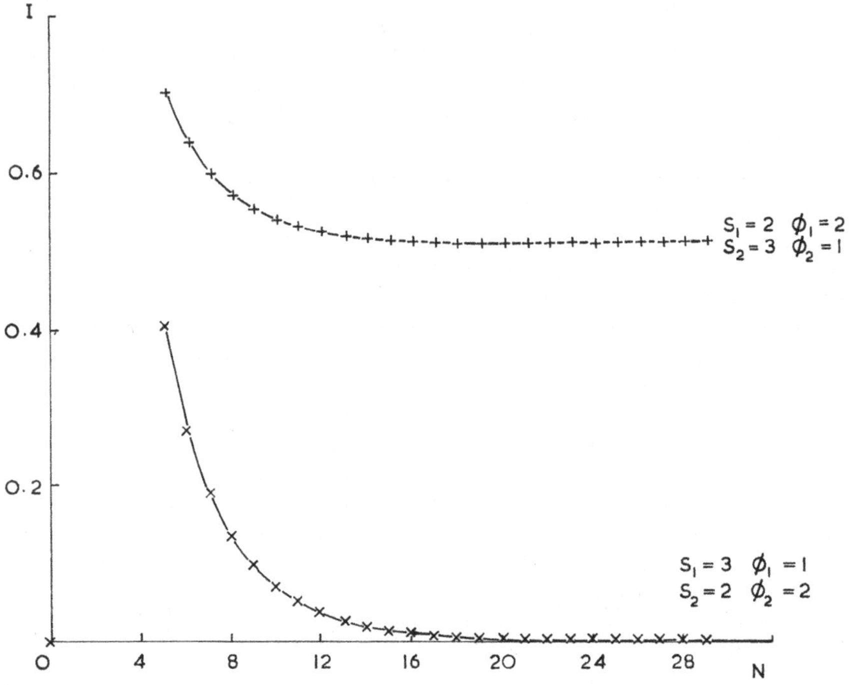

Fig. 2.3 The dependence of the average number of machines idle on the number of spare components

Abbreviating $\pi(n, N-n)$ to $\pi(n)$ we have that, when $N > s_1 + s_2$,

$$\pi(n) = \frac{B_N}{\phi_1^n n! \phi_2^{N-n} s_2! s_2^{N-n-s_2}} \qquad 0 \leq n \leq s_1$$

$$= \frac{B_N}{\phi_1^n s_1! s_1^{n-s_1} \phi_2^{N-n} s_2! s_2^{N-n-s_2}} \qquad s_1 \leq n \leq N - s_2$$

$$= \frac{B_N}{\phi_1^n s_1! s_1^{n-s_1} \phi_2^{N-n} (N-n)!} \qquad N - s_2 \leq n \leq N$$

Of course n is a birth and death process, and this fact could have been used to derive the above expressions. The normalizing constant B_N is determined by the identity $\sum \pi(n) = 1$, and elementary calculations show that

$$B_N = \frac{s_1! s_1^{-s_1} s_2! s_2^{N-s_2} \phi_2^N}{s_1! s_1^{-s_1} F(\rho s_1, s_1) + (\rho^{s_1+1} - \rho^{N-s_2})/(1-\rho) + s_2! s_2^{N-s_2} (\rho/s_2)^N F(s_2/\rho, s_2)}$$

where

$$\rho = \frac{\phi_2 s_2}{\phi_1 s_1}$$

and

$$F(x, r) = \sum_{n=0}^{r} \frac{x^n}{n!}$$

In equilibrium the average number of machines idle is

$$I = \sum_{n=0}^{s_1} (s_1 - n)\pi(n)$$

The dependence of I on N is illustrated in Fig. 2.3.

A mining operation. Consider a sequence of coal faces which are worked on in turn by a number of specialized machines. Examples of machines might be a cutting machine, a loading machine, and a roofing machine. Each machine proceeds to the next face after completing its task. We could regard the machines as queueing up at faces (Fig. 2.4a). However, the faces will

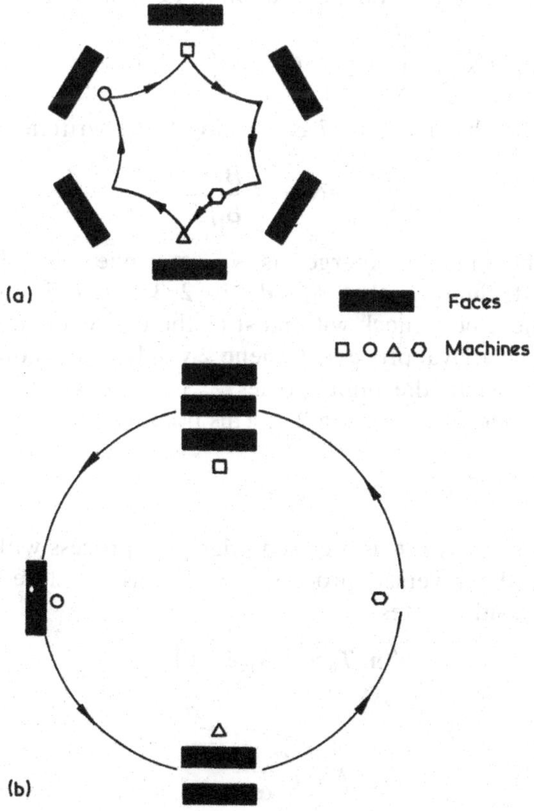

Fig. 2.4 A mining operation

usually be a more homogeneous group than the machines, and for this reason we shall regard the machines as fixed and the faces as queueing up for service from the machines (Fig. 2.4b). Suppose now that there are J machines and N faces and that the time taken by machine j to deal with a face is exponentially distributed with mean ϕ_j^{-1}, for $j=1,2,\ldots,J$. The system will then be a cyclic queue. If n_j is the number of faces queueing at machine j, then in equilibrium

$$\pi(n_1, n_2, \ldots, n_J) = \frac{B_N}{\prod_{j=1}^J \phi_j^{n_j}}$$

Note that the equilibrium probabilities do not depend upon the order in which machines work on faces. The normalizing constant is

$$B_N = \left[\sum_{\mathbf{n}\in\mathcal{S}} \phi_1^{-n_1}\phi_2^{-n_2}\cdots\phi_J^{-n_J}\right]^{-1}$$

and various quantities of interest depend upon it. For example the identity

$$\phi_j^{-1}B_{N-1}^{-1} = \left[\sum_{\mathbf{n}\in\mathcal{S}:n_j>0} \phi_1^{-n_1}\phi_2^{-n_2}\cdots\phi_J^{-n_J}\right]$$

allows the probability that machine j is working to be written as

$$\sum_{\mathbf{n}\in\mathcal{S}:n_j>0} \pi(\mathbf{n}) = \frac{B_N}{\phi_j B_{N-1}}$$

An interesting phenomenon emerges as $N\to\infty$ if one of the machines is slower than the rest. Suppose that $\phi_1 < \phi_j$, $j=2,3,\ldots,J$. Then as $N\to\infty$ queue 1 will become a bottleneck with most of the customers in the system waiting there, and the arrival process at queue 2 will become more and more like a Poisson process. In the limit queues $2,3,\ldots,J$ will behave as the series of queues considered in Section 2.2. This point is developed further in Exercise 2.4.5.

Exercises 2.3

1. Show that if the process $\mathbf{n}(t)$ is a closed migration process with transition rates (2.1) then the reversed process $\mathbf{n}(-t)$ is also a closed migration process, with transition rates

$$q'(\mathbf{n}, T_{jk}\mathbf{n}) = \lambda'_{jk}\phi_j(n_j)$$

where

$$\lambda'_{jk} = \frac{\alpha_k\lambda_{kj}}{\alpha_j}$$

Show that in equilibrium the probability flux that an individual moves from colony j to k is

$$\frac{\alpha_j \lambda_{jk} B_N}{B_{N-1}}$$

and deduce that in equilibrium the mean arrival rate at colony j, expression (2.7), is

$$\frac{\alpha_j \lambda_j B_N}{B_{N-1}}$$

2. Figure 2.3 suggests that I tends to a limit as $N \to \infty$. Prove this and show that the limit is zero if $\rho \geq 1$ and is $(1-\rho)s_1$ if $\rho \leq 1$.
3. Suppose that in the model of a mining operation $\phi_1 = \phi_2 = \cdots = \phi_J = \phi$. Show that in equilibrium the probability a given machine is operating is $N/(N+J-1)$ and that the average time for a machine to complete one cycle of faces is $(N+J-1)/\phi$.
4. Show that in the model of a mining operation the mean number of faces queueing for machine j can be written as

$$E(n_j) = \frac{\phi_j}{B_N}\left(\frac{\partial B_N}{\partial \phi_j}\right)$$

5. Consider a closed migration process in which each colony is a single-server queue. Suppose that a capacity constraint is put on each queue by the prohibition of any transition which would raise n_j above R_j, $j = 1, 2, \ldots, J$. Thus if $R = \sum_{j=1}^{J} R_j$ then we must have $R \geq N$. Suppose in addition

$$R - R_j < N \qquad j = 1, 2, \ldots, J$$

so that no queue can become empty. Show that if $m_j = R_j - n_j$ then $(m_1, m_2, \ldots, m_J)$ is a closed migration process, and hence deduce the equilibrium distribution for $(n_1, n_2, \ldots, n_J)$.
6. Show that the number of distinct states in a closed migration process is

$$\binom{N+J-1}{J-1}$$

Thus to calculate B_N directly as a sum of terms is impractical for even relatively small values of N and J. Fortunately there is an alternative. Define the generating functions

$$\Phi_j(z) = \sum_{n=0}^{\infty} \frac{(\alpha_j z)^n}{\prod_{r=1}^{n} \phi_j(r)}$$

$$B(z) = \sum_{N=0}^{\infty} \frac{z^N}{B_N}$$

Show that

$$B(z)=\prod_{j=1}^{J}\Phi_j(z)$$

Thus B_N can be calculated by multiplying together the functions $\Phi_j(z)$, $j=1,2,\ldots,J$, after they have each been truncated to the first $N+1$ terms. The number of steps required to do this is of order JN^2, and so this method is computationally much more efficient.

7. The generating function method readily yields marginal distributions. If β_n is the coefficient of z^N in

$$\frac{(\alpha_k z)^n}{\prod_{r=1}^{n}\phi_k(r)}\prod_{j\neq k}\Phi_j(z)$$

show that the probability colony k contains n individuals is $\beta_n B_N$.

8. In special cases the amount of computation required by the generating function method can be reduced further. If each colony is a single-server queue show that the form of the functions $\Phi_j(z)$, $j=1,2,\ldots,J$, allows B_N to be calculated in order JN steps. Show also that the probability queue k contains n or more customers is

$$\frac{\alpha_k^n B_N}{\phi_k^n B_{N-n}}$$

2.4 OPEN MIGRATION PROCESSES

In this section we shall again consider a set of J colonies but we shall allow individuals to enter and leave the system as well as to move between colonies. We will require the operators $T_{j\cdot}$ and $T_{\cdot k}$ defined as follows:

$$T_{j\cdot}(n_1,n_2,\ldots,n_j,\ldots,n_J)=(n_1,n_2,\ldots,n_j-1,\ldots,n_J)$$

$$T_{\cdot k}(n_1,n_2,\ldots,n_k,\ldots,n_J)=(n_1,n_2,\ldots,n_k+1,\ldots,n_J)$$

Thus $T_{j\cdot}$ removes an individual from colony j and $T_{\cdot k}$ introduces one at colony k. We shall study $\mathbf{n}$ under the assumption that it is a Markov process with transition rates given by

$$\begin{aligned} q(\mathbf{n},T_{jk}\mathbf{n})&=\lambda_{jk}\phi_j(n_j)\\ q(\mathbf{n},T_{j\cdot}\mathbf{n})&=\mu_j\phi_j(n_j)\\ q(\mathbf{n},T_{\cdot k}\mathbf{n})&=\nu_k \end{aligned} \tag{2.8}$$

where $\phi_j(0)=0$. We shall require that $\phi_j(n)>0$ if $n>0$ and that the parameters λ_{jk}, μ_j, and ν_k allow an individual to reach any colony from outside the system and to leave the system from any colony, either directly or indirectly via a chain of other colonies. Under these conditions the process **n** is irreducible within the state space $\mathbb{N}^J$, and we shall call it an open migration process.

The major difference between a closed and an open migration process is that in the latter individuals arrive at colonies from outside the system and individuals leaving colonies may well leave the system entirely. The transition rates (2.8) imply that arrivals at colony k from outside the system form a Poisson process of rate ν_k and that when an individual leaves colony j he will leave the system with probability μ_j/λ_j where

$$\lambda_j = \mu_j + \sum_k \lambda_{jk}$$

It is often convenient to scale the function ϕ_j so that $\lambda_j = 1$.

A series of simple queues (Fig. 2.2) is an example of an open migration process with $\phi_j(n) = \phi_j$, $n > 0$, where ϕ_j is the service rate at queue j, and with the only other non-zero parameters being $\nu_1 = \nu$ and $\lambda_{12} = \lambda_{23} = \cdots = \lambda_{J-1,J} = \mu_J = 1$. If we alter this system by setting $\mu_J = \lambda_{J1} = \frac{1}{2}$ we obtain the open migration process illustrated in Fig. 2.5, in which when a customer leaves queue J he returns to queue 1 with probability $\frac{1}{2}$ and leaves the system otherwise.

The conditions we have imposed on the parameters λ_{jk}, μ_j, and ν_k ensure that the equations

$$\alpha_j\left(\mu_j + \sum_k \lambda_{jk}\right) = \nu_j + \sum_k \alpha_k \lambda_{kj} \qquad j = 1, 2, \ldots, J \tag{2.9}$$

have a unique solution for $\alpha_1, \alpha_2, \ldots, \alpha_J$, which is positive (Exercise 2.4.1). We shall require as normalizing constants $b_1, b_2, \ldots, b_J$, where

$$b_j^{-1} = \sum_{n=0}^{\infty} \frac{\alpha_j^n}{\prod_{r=1}^{n} \phi_j(r)}$$

Let b_j be zero if the sum is infinite.

Theorem 2.4. *If $b_1, b_2, \ldots, b_J$ are all positive then the open migration process has an equilibrium distribution. In equilibrium $n_1, n_2, \ldots, n_J$ are independent and*

$$\pi_j(n_j) = b_j \frac{\alpha_j^{n_j}}{\prod_{r=1}^{n_j} \phi_j(r)} \qquad j = 1, 2, \ldots, J$$

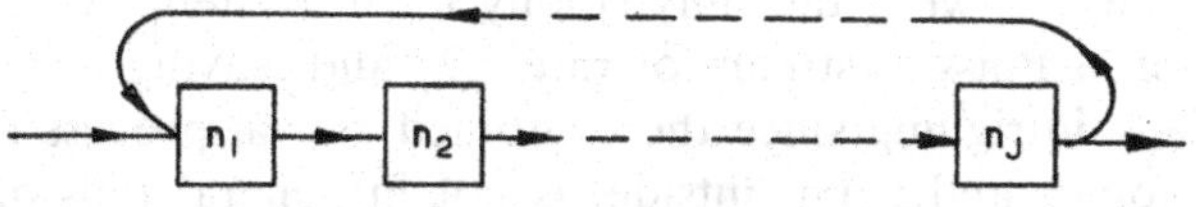

Fig. 2.5 An open migration process

Proof. The equilibrium equations are

$$\pi(\mathbf{n})\left[\sum_j q(\mathbf{n}, T_{j\cdot}\mathbf{n})+\sum_j\sum_k q(\mathbf{n}, T_{jk}\mathbf{n})+\sum_k q(\mathbf{n}, T_{\cdot k}\mathbf{n})\right]$$

$$=\sum_j \pi(T_{j\cdot}\mathbf{n})q(T_{j\cdot}\mathbf{n}, \mathbf{n})+\sum_j\sum_k \pi(T_{jk}\mathbf{n})q(T_{jk}\mathbf{n}, \mathbf{n})+\sum_k \pi(T_{\cdot k}\mathbf{n})q(T_{\cdot k}\mathbf{n}, \mathbf{n})$$

which will be satisfied if we can find a distribution $\pi(\mathbf{n})$ which satisfies the partial balance equations

$$\pi(\mathbf{n})\left[q(\mathbf{n}, T_{j\cdot}\mathbf{n})+\sum_k q(\mathbf{n}, T_{jk}\mathbf{n})\right]$$

$$=\pi(T_{j\cdot}\mathbf{n})q(T_{j\cdot}\mathbf{n}, \mathbf{n})+\sum_k \pi(T_{jk}\mathbf{n})q(T_{jk}\mathbf{n}, \mathbf{n}) \qquad j=1, 2, \ldots, J$$

and

$$\pi(\mathbf{n})\sum_k q(\mathbf{n}, T_{\cdot k}\mathbf{n})=\sum_k \pi(T_{\cdot k}\mathbf{n})q(T_{\cdot k}\mathbf{n}, \mathbf{n})$$

Substitution will verify that

$$\pi(\mathbf{n})=B\prod_{j=1}^{J}\frac{\alpha_j^{n_j}}{\prod_{r=1}^{n_j}\phi_j(r)} \tag{2.10}$$

satisfies the partial balance equations. For example the final partial balance equation reduces to, after substitution,

$$\sum_k \nu_k=\sum_k \alpha_k\mu_k$$

the truth of which is established by summing equations (2.9). Since $b_1, b_2, \ldots, b_J$ are positive the choice $B=b_1b_2\cdots b_J$ ensures that $\pi(\mathbf{n})$ sums to unity. Thus $\pi(\mathbf{n})$ is the equilibrium distribution and the independence of $n_1, n_2, \ldots, n_J$ follows from the fact that both $\pi(\mathbf{n})$ and the state space $\mathbb{N}^J$ have a product form.

The independence established in Theorem 2.4 is of the random variables $n_1, n_2, \ldots, n_J$ observed at a fixed point in time. Viewed as stochastic processes, defined for $t\in\mathbb{R}$, $n_1(t), n_2(t), \ldots, n_J(t)$ are clearly not independent.

It is interesting to note that the equilibrium distribution for colony j is just what it would be if it were the only colony in the system, with individuals arriving there in a Poisson stream of rate $\alpha_j\lambda_j$ and leaving at rate $\lambda_j\phi_j(n_j)$. This is especially intriguing since the combined arrival process at a colony, from other colonies and from outside, is not in general Poisson (Exercise 2.4.2).

If any of $b_1, b_2, \ldots, b_J$ are zero the process does not have an equilibrium distribution: there is a colony which individuals enter more quickly than they leave.

Observe that for the process to be reversible $\alpha_1, \alpha_2, \ldots, \alpha_J$ must satisfy

$$\begin{aligned} \alpha_j\lambda_{jk} &= \alpha_k\lambda_{kj} \\ \alpha_j\mu_j &= \nu_j \end{aligned} \tag{2.11}$$

Even when the process is not reversible the reversed process is of a similar form.

Theorem 2.5. *If* $\mathbf{n}(t)$ *is a stationary open migration process then so is the reversed process* $\mathbf{n}(-t)$.

Proof. Using Theorem 1.12 the transition rates of the reversed process $\mathbf{n}(-t)$ can be calculated from the rates (2.8) and the equilibrium distribution (2.10). For example

$$\begin{aligned} q'(\mathbf{n}, T_{jk}\mathbf{n}) &= \frac{\pi(T_{jk}\mathbf{n})q(T_{jk}\mathbf{n}, \mathbf{n})}{\pi(\mathbf{n})} \\ &= \lambda'_{jk}\phi_j(n_j) \end{aligned}$$

where

$$\lambda'_{jk} = \frac{\alpha_k\lambda_{kj}}{\alpha_j}$$

Similarly,

$$q'(\mathbf{n}, T_{j\cdot}\mathbf{n}) = \mu'_j\phi_j(n_j)$$

and

$$q'(\mathbf{n}, T_{\cdot k}\mathbf{n}) = \nu'_k$$

where

$$\mu'_j = \frac{\nu_j}{\alpha_j}$$

and

$$\nu'_k = \alpha_k\mu_k$$

Thus the reversed process is also an open migration process.

Call the points in time at which an individual leaves the system from colony j the exit process from colony j. By the departure process from colony j we shall mean the points in time at which an individual leaves colony j, either for another colony or to leave the system.

Corollary 2.6. *If $\mathbf{n}(t)$ is a stationary open migration process then the exit process from colony j is a Poisson process of rate $\alpha_j\mu_j$. Further, the exit processes from colonies $1, 2, \ldots, J$ are independent and $\mathbf{n}(t_0)$ is independent of the exit processes prior to time t_0.*

Proof. In the reversed process arrivals at colony j from outside the system form a Poisson process of rate $\nu_j' = \alpha_j\mu_j$. But these arrivals correspond precisely to departures from the system in the original process, and the result follows.

Neither the departure process from a colony nor the stream of customers moving from one colony to another is in general Poisson (Exercise 2.4.2). This again is intriguing. In the migration process illustrated in Fig. 2.5 an individual leaving colony J chooses at random and independently of everything that has gone before whether to leave the system or to return to colony 1. Yet while the departure process from colony J is not Poisson the exit process is. Note that the individual's decision on whether or not to return to colony 1 may be independent of past departures, but it is not independent of future departures.

Corollary 2.7. *Suppose that colony j in a stationary open migration process is a queue with s servers at which the queue discipline is first come first served. Let $\phi_j(n) = \phi_j \min(n, s)$ and $\lambda_j = 1$, so that service times are exponentially distributed with mean ϕ_j^{-1}. Then the waiting time of a customer at queue j has the same distribution as if queue j were an isolated $M/M/s$ queue with a Poisson arrival process of rate α_j.*

Proof. In a stationary open migration process the probability flux that a customer departs from queue j leaving n_j customers behind is $\pi_j(n_j + 1)\lambda_j\phi_j(n_j + 1)$. Thus if at time t a customer leaves queue j the probability there will be n_j customers left in queue j is

$$\frac{\pi_j(n_j+1)\phi_j(n_j+1)}{\sum_{r=0}^{\infty} \pi_j(r+1)\phi_j(r+1)} = \pi_j(n_j)$$

Consideration of the reversed process $\mathbf{n}(-t)$ shows that this is also the probability that a typical customer arriving at queue j finds n_j customers already there. But $\pi_j(n_j)$ is just what this probability would be if queue j were in isolation with customers arriving in a Poisson stream of rate α_j. The queue discipline ensures that the distribution of the waiting time of a customer is determined by the distribution of the number of customers he finds on his arrival, and the result follows.

Some of the simplest examples of open migration processes are those for which $\phi_j(n) = n$ for all j, i.e. linear migration processes. For these,

$b_1, b_2, \ldots, b_J$ will always be positive and so an equilibrium distribution will always exist. Indeed,

$$\pi_j(n_j) = e^{-\alpha_j} \frac{\alpha_j^{n_j}}{n_j!}$$

so that the number of individuals in colony j has a Poisson distribution. This result provides an alternative interpretation for the constants $\alpha_1, \alpha_2, \ldots, \alpha_J$; α_j is the expected number of individuals in colony j when individuals move independently with transition intensities λ_{jk}, μ_j, and ν_k.

Until now we have assumed that the number of colonies, J, is finite. In fact the proof of Theorem 2.4 goes through unchanged when J is infinite provided $B = b_1 b_2 \cdots$ is positive; note that when this is so the equilibrium distribution (2.10) assigns probability one to the countable set of states satisfying $\sum n_j < \infty$. In the following example we discuss a linear migration process with J infinite.

The family size process. Consider the following elaboration of the simple birth, death, and immigration process described in Section 1.3. Suppose that each immigrating individual has a distinguishing characteristic, such as a genetic type or a surname, which is passed on to all his descendants but which is not shared by any other individual. Thus at any point in time the population can be divided into distinct families, each of which consists of all those individuals alive with a given characteristic. Let n_j be the number of families of size j. Then the family size process $(n_1, n_2, \ldots)$ is a linear migration process with transition rates

$$q(\mathbf{n}, T_{j,j+1}\mathbf{n}) = j\lambda n_j \qquad j = 1, 2, \ldots$$

$$q(\mathbf{n}, T_{j,j-1}\mathbf{n}) = j\mu n_j \qquad j = 2, 3, \ldots$$

$$q(\mathbf{n}, T_{.1}\mathbf{n}) = \nu$$

$$q(\mathbf{n}, T_{1.}\mathbf{n}) = \mu n_1$$

Observe that a *family* is the basic unit which moves through the colonies of the system and that the movements of different families are independent. Equations (2.9) have the solution

$$\alpha_j = \frac{\nu}{\lambda j}\left(\frac{\lambda}{\mu}\right)^j$$

and the normalizing constant $B = \exp(-\sum \alpha_j)$ is positive provided $\lambda < \mu$, since then $\sum \alpha_j$ is finite. In equilibrium the process is reversible, the number of families of size j has a Poisson distribution with mean α_j, and the total number of families in the system has a Poisson distribution with mean

$$\sum \alpha_j = -\frac{\nu}{\lambda} \log\left(1 - \frac{\lambda}{\mu}\right)$$

Optimal allocation of effort. In this example we shall discuss an optimization application of Theorem 2.4. Consider an open migration process in which each colony is a single-server queue: suppose $\lambda_j = 1$, $\phi_j(n) = \phi_j$, $n > 0$, for $j = 1, 2, \ldots, J$. For equilibrium at each queue the service rate (or *effort*) ϕ_j must be greater than the mean arrival rate (or *demand*) α_j, and then

$$\pi_j(n_j) = \left(1 - \frac{\alpha_j}{\phi_j}\right)\left(\frac{\alpha_j}{\phi_j}\right)^{n_j}$$

Thus the mean number of customers present at queue j is $\alpha_j/(\phi_j - \alpha_j)$. Suppose now that we have control over the values of $\phi_1, \phi_2, \ldots, \phi_J$, subject to the constraint

$$\sum_j \phi_j = F$$

How should we choose $\phi_1, \phi_2, \ldots, \phi_J$ to minimize the mean number of customers present in the system? This problem can be readily solved using Lagrangian multipliers. Let

$$L = \sum_j \frac{\alpha_j}{\phi_j - \alpha_j} + y\left(\sum_j \phi_j - F\right)$$

Setting $\partial L/\partial \phi_j = 0$ we find that L is minimized by the choice

$$\phi_j = \alpha_j + \sqrt{\frac{\alpha_j}{y}}$$

Substituting this into the constraint shows that we should choose

$$\frac{1}{\sqrt{y}} = \frac{F - \sum_k \alpha_k}{\sum_k \sqrt{\alpha_k}}$$

Hence the optimal allocation is

$$\phi_j = \alpha_j + \frac{\sqrt{\alpha_j}}{\sum_k \sqrt{\alpha_k}}\left(F - \sum_k \alpha_k\right) \qquad j = 1, 2, \ldots, J$$

Thus the optimal allocation proceeds by first giving to each queue just enough to satisfy demand and then by allocating the surplus, $F - \sum_k \alpha_k$, in proportion to the square roots of the demands. This result is mildly surprising; we might have thought that effort would be allocated in proportion to demand. Relative to this allocation the optimal allocation concentrates less effort on those queues with high demands.

Little's result (1.12) shows that the optimal allocation also minimizes the mean period spent in the system by a customer.

A further discussion is contained in Section 4.1.

Exercises 2.4

1. By considering a Markov process with $J+1$ states and transition rates

$$q(j,k)=\lambda_{jk} \qquad j,k=1,2,\ldots,J$$
$$q(j,0)=\mu_j \qquad j=1,2,\ldots,J$$
$$q(0,k)=\nu_k \qquad k=1,2,\ldots,J$$

show that equations (2.9) have a unique solution and that this solution is positive. Show that $\alpha_j\lambda_j/\sum \nu_k$ is the expected number of times the jump chain of this process visits state j between successive visits to state 0. Deduce that in an open migration process $\alpha_j\lambda_j$ is the mean arrival rate at colony j, counting arrivals from outside the system and from other colonies. Obtain the same result by calculating the probability flux that an individual leaves colony j.

2. Consider the open migration process illustrated in Fig. 2.5 with $J=2$, $\phi_j(n)=n$, $j=1,2$, $\mu_1=\nu_2=0$, $\lambda_{12}=\lambda_{21}=\lambda$, $\nu_1=\nu$, and $\mu_2=\mu$. Show that the arrival process at colony 1, counting arrivals from outside the system and from colony 2, comprises a Poisson process of rate ν together with for each point of this process a string of further points, where the number of further points in each string is geometrically distributed with mean λ/μ and the interval between points in the same string has mean $2\lambda^{-1}$. Suppose now that ν is small and λ large. Show that the arrival process at colony 1 is not Poisson. Deduce that the departure process from colony 2 is not Poisson. Show that the points in time at which individuals move from colony 2 to colony 1 do not form a Poisson process.

3. Consider an open migration process. If it is not possible for an individual in colony k ever to reach colony j show that the stream of individuals moving from colony j to colony k is Poisson.

4. Consider an open migration process in which an individual can never visit a colony more than once, and the graph G, with an edge joining nodes j and k if either λ_{jk} or λ_{kj} is positive, is a tree. If each queue is a first come first served single-server queue show that the waiting times of a customer at the queues he visits are independent. Note that the conditions ensure that customers cannot overtake one another. Using Exercise 2.2.4 show that the conclusion remains valid if some queues have more than one server provided these queues are such that a customer can only visit them immediately on entering or immediately prior to leaving the system.

5. Consider a stationary closed migration process with colonies labelled $0,1,2,\ldots,J$ and with $\lambda_{j0}=\mu_j$, $\lambda_{0j}=\nu_j$, for $j=1,2,\ldots,J$, and $\phi_0(n)=1$, $n>0$. Let $\alpha_1,\alpha_2,\ldots,\alpha_J$ be the solution to equations (2.9) and suppose the constants $b_1,b_2,\ldots,b_J$ calculated from this solution are all positive.

Show that if the number of individuals in the system is increased towards infinity the behaviour of colonies $1, 2, \ldots, J$ approaches that of an open migration process.

6. The equilibrium distribution obtained for the family size process should be consistent with that found for the population size in Section 1.3. Establish this directly by showing that if

$$M = \sum_{j=1}^{\infty} jn_j$$

where $n_1, n_2, \ldots$ are independent random variables, n_j Poisson with mean

$$\frac{\nu}{\lambda j}\left(\frac{\lambda}{\mu}\right)^j$$

then M has the negative binomial distribution (1.14).

7. In the family size process show that the total number of individuals M and the total number of families

$$N = \sum_{j=1}^{\infty} n_j$$

satisfy the relations

$$E(N) = \frac{\nu}{\lambda} \log\left(1 + \frac{\lambda}{\nu} E(M)\right)$$

and

$$\operatorname{cov}(M, N) = \frac{\nu}{\mu - \lambda}$$

8. Consider the family size process. Show that if an individual is the only member of his family then he is an immigrant who has not yet given birth with probability $\mu/(\mu + \lambda)$.
9. In the family size process show that the points in time at which a family becomes extinct form a Poisson process. Show that this remains true even when the model is amended to allow the birth of twins.

CHAPTER 3

Queueing Networks

In the previous chapter some simple examples of queueing networks were introduced. This chapter will continue the discussion of queueing networks, but within a more general framework.

3.1 GENERAL CUSTOMER ROUTES

Consider the queueing network illustrated in Fig. 3.1. In this network there are five simple queues, and customers can enter the system at queues 1 or 2, arrivals at these queues forming two independent Poisson processes. Customers follow the route through queues 1, 3, and 4 or the route through queues 2, 3, and 5 before leaving the system. This might be a model of a manufacturing job-shop with customers representing items of work which require to be processed at a sequence of machines. This network *cannot* be represented by a migration process. The difficulty is that a customer leaving queue 3 does not choose at random between queues 4 and 5: he moves to queue 4 if he has previously been to queue 1. In a migration process the past route of a customer in a given queue is of no use in predicting his future route, and in this sense the customers in a queue are homogeneous. In this section we shall see that by dividing customers into different types we can deal with networks such as the one illustrated in Fig. 3.1.

Suppose that there are I different customer types and that a customer's type determines his route through the J queues of the system. More specifically, suppose that customers of type i $(i = 1, 2, \ldots, I)$ enter the system in a Poisson stream at rate $\nu(i)$ and pass through the sequence of queues

$$r(i, 1), r(i, 2), \ldots, r(i, S(i))$$

before leaving the system. Thus the queue which a customer of type i visits at stage s $(s = 1, 2, \ldots, S(i))$ of his route is queue $r(i, s)$. Note that the route of a customer may require him to visit the same queue more than once. For simplicity we shall not allow two successive stages of a customer's route to be identical. We shall assume that the I Poisson arrival streams are independent. It is not essential that I be finite, but we shall require that $\sum \nu(i)$ be finite.

By using more than one customer type we can represent the behaviour of a customer whose future route depends stochastically upon his past route: we simply use a different type for each possible route. Consider, for

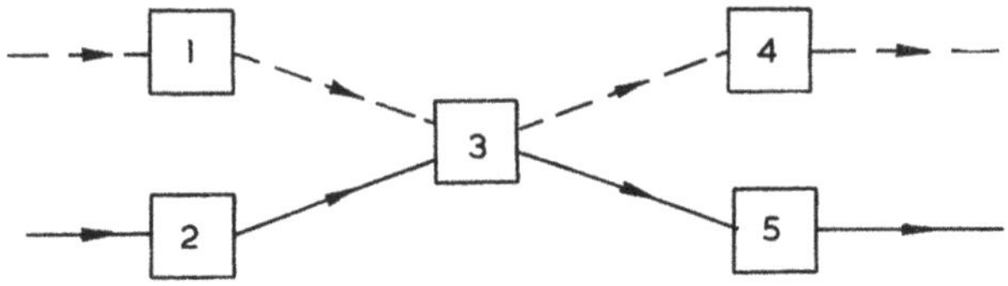

Fig. 3.1 A job-shop model

example, the network illustrated in Fig. 3.2. Suppose this differs from the network of Fig. 3.1 in that customers who have been through queue 2 are, after leaving queue 3, equally likely to move to queue 4 or queue 5. We require three customer types to model this network: customers of types 1, 2, and 3 follow the routes $1 \to 3 \to 4$, $2 \to 3 \to 4$, and $2 \to 3 \to 5$ respectively, and the arrival rates $\nu(2)$ and $\nu(3)$ are equal.

The above method can deal with the random routes which arise in an open migration process, but it will be more cumbersome than the approach of the previous chapter if the migration process allows a customer to visit the same queue more than once (Exercise 3.1.2). The advantage of the above method is that it allows much more general routing schemes than can arise in a migration process. To give two further examples, it can deal with a system in which a customer visits each queue exactly once, but in a random order, or a system in which each customer visits a certain queue exactly twice.

We have described how customers move between queues: we must now describe how the queues themselves operate. This is rather more complicated than it was for a migration process, since within each queue we must now keep track of the different types of customer. We shall suppose that the customers in each queue are ordered: thus queue j $(j = 1, 2, \ldots, J)$ will contain customers in positions $1, 2, \ldots, n_j$, where n_j is the total number of customers in queue j. Assume queue j operates in the following manner:

(i) Each customer requires an amount of service which is a random variable exponentially distributed with unit mean.

(ii) A total service effort is supplied at the rate $\phi_j(n_j)$.

(iii) A proportion $\gamma_j(l, n_j)$ of this effort is directed to the customer in position l $(l = 1, 2, \ldots, n_j)$; when this customer leaves the queue, his service

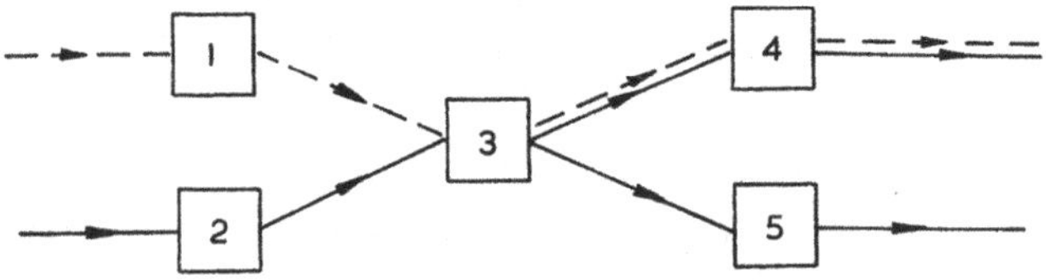

Fig. 3.2 Random routes

completed, customers in positions $l+1, l+2, \ldots, n_j$ move to positions $l, l+1, \ldots, n_j-1$ respectively.

(iv) When a customer arrives at queue j he moves into position l $(l=1, 2, \ldots, n_j+1)$ with probability $\delta_j(l, n_j+1)$; customers previously in positions $l, l+1, \ldots, n_j$ move to positions $l+1, l+2, \ldots, n_j+1$ respectively.

Of course

$$\sum_{l=1}^{n} \gamma_j(l, n) = 1$$

$$\sum_{l=1}^{n} \delta_j(l, n) = 1$$

and we shall insist that $\phi_j(n)>0$ if $n>0$. Call the amount of service a customer requires at a queue his *service requirement.* We shall assume that all service requirements, even of the same customer at different queues, are independent of each other and of the times at which customers enter the system. The way in which a customer's service requirement is satisfied can be visualized as follows. While the queue contains n_j customers, with him in position l, he receives service effort at the rate $\phi_j(n_j)\gamma_j(l, n_j)$ per unit time. When the amount of service effort he has received reaches his service requirement he leaves the queue. Since service requirements are exponentially distributed with unit mean, if queue j contains n_j customers then the probability intensity that the customer in position l leaves is $\phi_j(n_j)\gamma_j(l, n_j)$.

To illustrate the behaviour which can be allowed, if $\phi_j(n) = \lambda_j \min(K, n)$

$$\gamma_j(l, n) = \begin{cases} \dfrac{1}{n} & l=1, 2, \ldots, n;\; n=1, 2, \ldots, K \\ \dfrac{1}{K} & l=1, 2, \ldots, K;\; n=K+1, K+2, \ldots \\ 0 & \text{otherwise} \end{cases}$$

$$\delta_j(l, n) = \begin{cases} 1 & l=n \\ 0 & \text{otherwise} \end{cases}$$

then queue j behaves as a K-server queue in which customers have their service commenced in the order of their arrival and each customer has an exponentially distributed service time with mean λ_j^{-1}. In this example the service time of a customer can be identified with his service requirement, but this will not always be so. By varying ϕ_j we can allow the servers to work faster when the queue is large. By varying δ_j we can alter the queue discipline, making it, for example, last come first served or service in random order. A more subtle use of ϕ_j and γ_j will let a waiting customer defect at a rate depending upon his position in the queue. Note, however, that we cannot model a priority discipline based upon the type of a customer; nor can we allow a customer's service time to depend upon his service time at previous queues.

Let $t_j(l)$ be the type of the customer in position l in queue j and let $s_j(l)$ be the stage along his route that this customer has reached. We shall call $c_j(l) = (t_j(l), s_j(l))$ the class of this customer; if he can visit queue j more than once his class will contain more information than his type. The vector

$$\mathbf{c}_j = (c_j(1), c_j(2), \ldots, c_j(n_j))$$

describes the state of queue j and

$$\mathbf{C} = (\mathbf{c}_1, \mathbf{c}_2, \ldots, \mathbf{c}_J)$$

is a Markov process representing the state of the system.

What are the transition rates of the process $\mathbf{C}$? If the customer in position l in queue j is at the last stage of his route then a possible event is that this customer may leave the system. Let $T_{jl}.\mathbf{C}$ be the state of the process after this event. The probability intensity of the event is

$$q(\mathbf{C}, l, \cdot, T_{jl}.\mathbf{C}) = \phi_j(n_j)\gamma_j(l, n_j) \tag{3.1}$$

It may be that $T_{jl}.\mathbf{C} = T_{jg}.\mathbf{C}$ for $l \neq g$, for example if all the customers in queue j are of the same type. The transition rate from the state $\mathbf{C}$ to the state $T_{jl}.\mathbf{C}$ is given by

$$q(\mathbf{C}, T_{jl}.\mathbf{C}) = \sum_g q(\mathbf{C}, g, \cdot, T_{jg}.\mathbf{C}) \tag{3.2}$$

where the summation runs over g such that $T_{jl}.\mathbf{C} = T_{jg}.\mathbf{C}$. If the customer in position l in queue j is not at the last stage of his route then let $k = r(t_j(l), s_j(l)+1)$ be the next queue he will visit. In this case a possible event is that this customer may leave queue j and move into position m in queue k. Let $T_{jlm}\mathbf{C}$ be the state of the process after this event. The probability intensity of the event is

$$q(\mathbf{C}, l, m, T_{jlm}\mathbf{C}) = \phi_j(n_j)\gamma_j(l, n_j)\delta_k(m, n_k+1) \tag{3.3}$$

The transition rate from the state $\mathbf{C}$ to the state $T_{jlm}\mathbf{C}$ is given by

$$q(\mathbf{C}, T_{jlm}\mathbf{C}) = \sum_g \sum_h q(\mathbf{C}, g, h, T_{jgh}\mathbf{C}) \tag{3.4}$$

where the summation runs over g and h such that $T_{jlm}\mathbf{C} = T_{jgh}\mathbf{C}$. Another possible event is that a customer of type i may enter the system and move into position m in queue k, where $k = r(i, 1)$. Let $T^{im}\mathbf{C}$ be the state of the process after this event. The probability intensity of the event is

$$q(\mathbf{C}, \cdot, m, T^{im}\mathbf{C}) = \nu(i)\delta_k(m, n_k+1) \tag{3.5}$$

The transition rate from the state $\mathbf{C}$ to the state $T^{im}\mathbf{C}$ is given by

$$q(\mathbf{C}, T^{im}\mathbf{C}) = \sum_h q(\mathbf{C}, \cdot, h, T^{ih}\mathbf{C}) \tag{3.6}$$

where the summation runs over h such that $T^{im}\mathbf{C} = T^{ih}\mathbf{C}$.

Of course for a given state $\mathbf{C}$ it would not be appropriate to apply certain of the T operators defined above. However, we can say that any non-zero transition rate of the process $\mathbf{C}$ is of the form (3.2), (3.4), or (3.6).

Let

$$\alpha_j(i, s) = \begin{cases} \nu(i) & \text{if } r(i, s) = j \\ 0 & \text{otherwise} \end{cases}$$

and let

$$a_j = \sum_{i=1}^{I} \sum_{s=1}^{S(i)} \alpha_j(i, s)$$

If the system is in equilibrium then a_j will be the average number of customers arriving at queue j per unit time. Let

$$b_j^{-1} = \sum_{n=0}^{\infty} \frac{a_j^n}{\prod_{l=1}^{n} \phi_j(l)}$$

We shall assume that none of $b_1, b_2, \ldots, b_J$ is zero. This condition is imposed to ensure that an equilibrium distribution for the system exists, and if it is not satisfied at least one queue will be unable to cope with the number of customers arriving at it. Define

$$\pi_j(\mathbf{c}_j) = b_j \prod_{l=1}^{n_j} \frac{\alpha_j(t_j(l), s_j(l))}{\phi_j(l)}$$

Theorem 3.1. *The equilibrium distribution for the open network of queues described above is*

$$\pi(\mathbf{C}) = \prod_{j=1}^{J} \pi_j(\mathbf{c}_j)$$

Proof. First notice that $\pi(\mathbf{C})$ sums to unity, by the definition of the constants $b_1, b_2, \ldots, b_J$.

What might the Markov process $\mathbf{C}(t)$ look like if we reversed the direction of time? One possibility is that customers of type i might enter the system in a Poisson stream at rate $\nu(i)$ and pass through the sequence of queues

$$r(i, S(i)), r(i, S(i)-1), \ldots, r(i, 1)$$

before leaving the system, and that the queues of the system might behave as before but with the functions γ_j and δ_j interchanged. The reversed process $\mathbf{C}(-t)$ would then be of the same form as $\mathbf{C}(t)$, but with different parameters. With this in mind define, corresponding to the probability intensities (3.1), (3.3), and (3.5),

$$q'(T_{jl}.\mathbf{C}, \cdot, l, \mathbf{C}) = \nu(i)\gamma_j(l, n_j) \qquad \text{where } i = t_j(l)$$

$$q'(T_{jlm}\mathbf{C}, m, l, \mathbf{C}) = \phi_k(n_k+1)\delta_k(m, n_k+1)\gamma_j(l, n_j) \quad \text{where } k = r(t_j(l), s_j(l)+1)$$

$$q'(T^{im}\mathbf{C}, m, \cdot, \mathbf{C}) = \phi_k(n_k+1)\delta_k(m, n_k+1) \qquad \text{where } k = r(i, 1)$$

Similarly define the transition rates $q'(\mathbf{C}, \mathbf{D})$ by analogy with the definition of the transition rates $q(\mathbf{C}, \mathbf{D})$. By substituting the proposed form for $\pi(\mathbf{C})$ we see that

$$\pi(\mathbf{C})q(\mathbf{C}, l, m, T_{jlm}\mathbf{C}) = \pi(T_{jlm}\mathbf{C})q'(T_{jlm}\mathbf{C}, m, l, \mathbf{C})$$

Thus, by summation,

$$\pi(\mathbf{C})q(\mathbf{C}, T_{jlm}\mathbf{C}) = \pi(T_{jlm}\mathbf{C})q'(T_{jlm}\mathbf{C}, \mathbf{C})$$

In this way we can establish that for all $\mathbf{C}$ and $\mathbf{D}$,

$$\pi(\mathbf{C})q(\mathbf{C}, \mathbf{D}) = \pi(\mathbf{D})q'(\mathbf{D}, \mathbf{C})$$

We also find that

$$q(\mathbf{C}) = q'(\mathbf{C}) = \sum_{j=1}^{J} \phi_j(n_j) + \sum_{i=1}^{I} \nu(i)$$

Hence Theorem 1.13 allows us to deduce that $\pi(\mathbf{C})$ is the equilibrium distribution for the process $\mathbf{C}(t)$, which completes the proof of the present result.

We can also deduce from Theorem 1.13 that $\mathbf{C}(-t)$ does indeed take the form suggested; thus we obtain the following result.

Theorem 3.2. *If $\mathbf{C}(t)$ is a stationary open network of queues of the form described in this section then so is the reversed process $\mathbf{C}(-t)$.*

Theorems 3.1 and 3.2 parallel Theorems 2.4 and 2.5, and as in Chapter 2 there are some immediate consequences. Theorem 3.2 has the following corollary.

Corollary 3.3. *In equilibrium customers of type i ($i = 1, 2, \ldots, I$) leave the system in a Poisson stream at rate $\nu(i)$. These I Poisson streams are independent, and $\mathbf{C}(t_0)$ is independent of departures from the system prior to time t_0.*

If $\mathbf{c}_1, \mathbf{c}_2, \ldots, \mathbf{c}_J$ are possible states for the queues $1, 2, \ldots, J$ then $\mathbf{C} = (\mathbf{c}_1, \mathbf{c}_2, \ldots, \mathbf{c}_J)$ is a possible state for the system. This implies that the state space $\mathcal{S}$ has a product form and hence we can deduce from Theorem 3.1 that in equilibrium $\mathbf{c}_1, \mathbf{c}_2, \ldots, \mathbf{c}_J$ are independent.

Corollary 3.4. *In equilibrium the state of queue j is independent of the state of the rest of the system and is $\mathbf{c}_j$ with probability $\pi_j(\mathbf{c}_j)$. The probability that queue j contains n customers is*

$$P(n_j = n) = b_j \frac{a_j^n}{\prod_{l=1}^n \phi_j(l)} \tag{3.7}$$

If a customer is in position l in queue j then the probability that he is a type i customer at stage s of his route is $\alpha_j(i, s)/a_j$.

Equation (3.7) is exactly the expression we obtain if queue j is a single queue with customers arriving in a Poisson stream at rate a_j. Note, however, that in general arrivals at queue j do not form a Poisson process (cf. Exercise 2.4.2).

Corollary 3.5. *When a customer of type i reaches queue j at stage s of his route the probability that he finds queue j in state $\mathbf{c}_j$ is $\pi_j(\mathbf{c}_j)$. The probability that he finds n customers in queue j is given by expression (3.7).*

Proof. If $s = 1$ the result follows immediately from the fact that the arrival process of type i customers at the first queue on their route is Poisson. For $s > 1$ the proof proceeds along the same lines as the proof of Corollary 2.7. In equilibrium the probability flux that a customer of type i will depart from queue j after a given stage of his route and that the queue will be left in state $\mathbf{c}_j$ with n_j customers is

$$\sum_{l=1}^{n_j+1} \pi_j(\mathbf{c}_j) \frac{\nu(i)}{\phi_j(n_j+1)} \phi_j(n_j+1)\delta_j(l, n_j+1) = \nu(i)\pi_j(\mathbf{c}_j)$$

Thus if a customer of type i has just left queue j after a given stage of his route, the probability that he has left queue j in state $\mathbf{c}_j$ is $\pi_j(\mathbf{c}_j)$. Consideration of the reversed process now establishes the desired result.

If queue j is a first come first served K-server queue then Corollary 3.5 shows that the waiting time of a customer at this queue has the same distribution as if queue j were an isolated $M/M/K$ queue with a Poisson arrival process of rate a_j. Note that the waiting time of a customer at queue j will not in general be independent of his experience elsewhere in the network (cf. Exercise 2.2.5).

Exercises 3.1

1. Show that for the network illustrated in Fig. 3.1 the process $(n_1, n_2, \ldots, n_5)$ is not Markov.

2. Consider an open migration process with transition rates (2.8) where $\lambda_j = 1$, $j = 1, 2, \ldots, J$. Show that the process can be regarded as a queueing network with customers whose route will be $r(i, 1), r(i, 2), \ldots, r(i, S(i))$ arriving at rate

$$\nu_{r(i,1)}\lambda_{r(i,1),r(i,2)}\lambda_{r(i,2),r(i,3)}\cdots\lambda_{r(i,S(i)-1),r(i,S(i))}\mu_{r(i,S(i))}.$$

Observe that an infinite number of types will be required if a customer can visit the same queue more than once. Show that the quantities $a_1, a_2, \ldots, a_J$ calculated from the queueing network parameters are equal to the quantities $\alpha_1, \alpha_2, \ldots, \alpha_J$ determined by equations (2.9) from the migration process parameters.

3. Suppose that in the description given of a K-server queue the function δ_j is altered to

$$\delta_j(l, n) = \begin{cases} 1 & l = n;\ n = 1, 2, \ldots, K \\ 1 & l = K+1;\ n = K+1, K+2, \ldots \\ 0 & \text{otherwise} \end{cases}$$

Show that the resulting queue discipline is last come first served without preemption. If

$$\delta_j(l, n) = \frac{1}{n-K} \qquad l = K+1, K+2, \ldots, n;\ n = K+1, K+2, \ldots$$

show that the queue discipline is service in random order (i.e. that the queue is equivalent to one in which when a customer leaves the queue the next customer to be served is chosen at random from amongst those whose service has not yet commenced).

4. Suppose that $\phi_j(n) = \phi_j$ for all $n > 0$, $j = 1, 2, \ldots, J$, so that each queue is a single-server queue. Observe that n_j is a geometric random variable with mean $a_j/(\phi_j - a_j)$. Show that the number of type i customers at stage s of their route is also a geometric random variable, with mean $\nu(i)/(\phi_j - a_j)$ where $j = r(i, s)$. Observe that for differing values of i and s giving rise to the same value of j these random variables are dependent. Deduce from Little's result (1.12) that the mean time it takes a type i customer to pass through the system is

$$\sum_{s=1}^{S(i)} [\phi_{r(i,s)} - a_{r(i,s)}]^{-1}$$

5. Show that the restriction not allowing two successive stages of a customer's route to be identical can be removed.

6. The requirement that $\phi_j(n) > 0$ if $n > 0$ can be relaxed. Find the equilibrium distribution for a system in which

$$\phi_j(K) = 0$$

$$\phi_j(n) > 0 \qquad n > K$$

This form of the function ϕ_j would correspond to the servers at queue j only operating when more than K customers are present.

7. Show that the results of this section are unaltered if the functions $\gamma_j(l, n_j)$, $\delta_j(l, n_j)$ are replaced by functions $\gamma_j(l, \mathbf{c}_j)$, $\delta_j(l, \mathbf{c}_j)$, provided the functions γ_j and δ_j are invariant under permutations of $\mathbf{c}_j = (c_j(1), c_j(2), \ldots, c_j(n_j))$ and
$$\sum_{l=1}^{n_j} \gamma_j(l, \mathbf{c}_j) = \sum_{l=1}^{n_j} \delta_j(l, \mathbf{c}_j) = 1$$

3.2 OPEN NETWORKS OF QUASI-REVERSIBLE QUEUES

The routing mechanism introduced in the previous section is general enough for most purposes, but the queue described there is fairly limited in scope. In this section we shall show that essentially the same results can be obtained for any network of queues provided the queues have a certain important characteristic.

To define this characteristic we shall begin by considering a single isolated queue. We shall make quite weak assumptions about the nature of this queue; it could perhaps be visualized as a black box with a stream of customers entering the box and a further stream of customers leaving the box. Assume that every customer entering the queue leaves it but, for simplicity, not immediately. Assume also that at no point in time does more than one customer enter or leave the queue. Further assume that each customer has a class c chosen from a countable set $\mathscr{C}$ and that customers do not change class as they pass through the queue. Often the class of a customer will convey information about him; later we shall use it to provide an indication of his past and future route in a network and his service requirements at the various queues of the network. Suppose there is associated with the queue a Markov process $\mathbf{x}(t)$, which we shall call the state of the queue at time t. Assume that the state of the queue contains enough information for us to deduce how many customers of each class there are in the queue. Often the state will contain further information concerning, for example, the arrangement of customers within the queue or the amount of service still required by each customer. From now on we shall identify the queue with the Markov process $\mathbf{x}(t)$ giving its state. Observe that from a realization of the process $\mathbf{x}(t)$, $-\infty < t < \infty$, we can construct the arrival and departure processes of customers of class c, since such arrivals and departures are signalled by changes in the number of customers of class c in the queue.

Definition

A queue is *quasi-reversible* if its state $\mathbf{x}(t)$ is a stationary Markov process with the property that the state of the queue at time t_0, $\mathbf{x}(t_0)$, is independent

of:
(i) the arrival times of class c customers, $c \in \mathscr{C}$, subsequent to time t_0;
(ii) the departure times of class c customers, $c \in \mathscr{C}$, prior to time t_0.

Theorem 3.6. *If a queue is quasi-reversible then:*
(i) *arrival times of class c customers, for $c \in \mathscr{C}$, form independent Poisson processes;*
(ii) *departure times of class c customers, for $c \in \mathscr{C}$, form independent Poisson processes.*

Proof. Let $\mathscr{S}(c, \mathbf{x})$ be the set of states in which the queue contains one more customer of class c than in state $\mathbf{x}$, with the same numbers of customers of other classes. Thus a transition from the state $\mathbf{x}$ to a state $\mathbf{x}' \in \mathscr{S}(c, \mathbf{x})$ indicates the arrival of a customer of class c. Since the queue is quasi-reversible the probability a customer of class c arrives in the interval $(t_0, t_0+\delta t)$ is independent of the state $\mathbf{x}(t_0)$. Hence the probability intensity that a customer of class c arrives when the state is $\mathbf{x}$ depends only on c and not on $\mathbf{x}$; call it

$$\alpha(c) = \sum_{\mathbf{x}' \in \mathscr{S}(c,\mathbf{x})} q(\mathbf{x}, \mathbf{x}') \tag{3.8}$$

Since $\mathbf{x}(t)$ is a Markov process the realization $\mathbf{x}(t)$, $-\infty < t \leq t_0$, contains no more information than does $\mathbf{x}(t_0)$ about whether or not a class c customer will arrive in the interval $(t_0, t_0+\delta t)$. But this realization gives the arrival times of class c customers, for $c \in \mathscr{C}$, prior to time t_0. Hence the probability intensity that a customer of class c will arrive is $\alpha(c)$, even given all prior arrival times of class c customers, for $c \in \mathscr{C}$. Hence arrival times of class c customers, for $c \in \mathscr{C}$, form independent Poisson processes.

Consider now the reversed process $\mathbf{x}(-t)$. This can also be regarded as a queue: again transitions from the state $\mathbf{x}$ to a state $\mathbf{x}' \in \mathscr{S}(c, \mathbf{x})$ indicate the arrival of a customer of class c and transitions to the state $\mathbf{x}$ from a state $\mathbf{x}' \in \mathscr{S}(c, \mathbf{x})$ indicate the departure of a customer of class c. Observe that arrivals at the reversed queue $\mathbf{x}(-t)$ subsequent to time $-t_0$ correspond to departures from the queue $\mathbf{x}(t)$ prior to time t_0. Similarly, departures from the reversed queue $\mathbf{x}(-t)$ prior to time $-t_0$ correspond to arrivals at the queue $\mathbf{x}(t)$ subsequent to time t_0. Since the queue $\mathbf{x}(t)$ is quasi-reversible it therefore follows that the reversed queue $\mathbf{x}(-t)$ is also quasi-reversible. Thus at the reversed queue $\mathbf{x}(-t)$ arrival times of class c customers, for $c \in \mathscr{C}$, form independent Poisson processes. Thus at the queue $\mathbf{x}(t)$ departure times of class c customers, for $c \in \mathscr{C}$, form independent Poisson processes.

Although conclusions (i) and (ii) of Theorem 3.6 are the most obvious features of a quasi-reversible queue they cannot be taken as the definition of quasi-reversibility. These conclusions include no mention of the Markov process $\mathbf{x}(t)$ defining the state of the queue, and it is possible to construct systems satisfying conclusions (i) and (ii) which are not quasi-reversible (Exercises 3.2.2 and 3.2.3).

It usually follows from the definition of the process $\mathbf{x}(t)$ that $\mathbf{x}(t_0)$ is independent of subsequent arrivals. Often the form of the reversed process $\mathbf{x}(-t)$ allows us to deduce that $\mathbf{x}(t_0)$ is also independent of prior departures and hence that the queue is quasi-reversible. An example of a quasi-reversible queue is an $M/M/1$ queue with one class of customer and with $\mathbf{x}(t)$ the number in the queue at time t. Theorem 2.1 establishes that this queue is quasi-reversible. More generally, if a queue has one class of customer, a Poisson arrival process, and the state of the queue is a reversible Markov process independent of future arrivals, then the queue will be quasi-reversible. More complicated examples are provided by the networks of the previous section. If the class of a customer is taken to be its type then Corollary 3.3 shows that a network, considered in its entirety as a single system, is a quasi-reversible queue. The special case in which the network consists of just one queue shows that a single queue of the form discussed in the previous section is quasi-reversible. Further examples of quasi-reversible queues will be discussed in the next section.

If $\pi(\mathbf{x})$ is the equilibrium distribution of the queue $\mathbf{x}(t)$ then the transition rates of the reversed queue $\mathbf{x}(-t)$ are given by

$$\pi(\mathbf{x})q'(\mathbf{x}, \mathbf{x}') = \pi(\mathbf{x}')q(\mathbf{x}', \mathbf{x}) \tag{3.9}$$

Departures of class c customers from the queue $\mathbf{x}(t)$ form a Poisson process; the rate of this process must be $\alpha(c)$ since this is the arrival rate and the queue is in equilibrium. Hence the arrival rate of class c customers at the reversed queue $\mathbf{x}(-t)$ is also $\alpha(c)$, and so

$$\alpha(c) = \sum_{\mathbf{x}' \in \mathcal{S}(c,\mathbf{x})} q'(\mathbf{x}, \mathbf{x}') \tag{3.10}$$

This is an important result; relations (3.8) and (3.10) characterize the property of quasi-reversibility for a stationary Markov process $\mathbf{x}(t)$. Using equations (3.8), (3.9), and (3.10) we can obtain the partial balance equations

$$\pi(\mathbf{x}) \sum_{\mathbf{x}' \in \mathcal{S}(c,\mathbf{x})} q(\mathbf{x}, \mathbf{x}') = \sum_{\mathbf{x}' \in \mathcal{S}(c,\mathbf{x})} \pi(\mathbf{x}')q(\mathbf{x}', \mathbf{x}) \tag{3.11}$$

Thus in equilibrium the probability flux out of a state due to a customer of class c arriving is equal to the probability flux into that same state due to a customer of class c departing. Since the probability flux that a customer of class c arrives at the queue is equal to the probability flux that a customer of class c departs from the queue, this shows that the distribution over states found by an arriving customer of class c is the same as that left behind by a departing customer of class c. If the process $\mathbf{x}(t)$ is reversible then the partial balance equations (3.11) are automatically satisfied; however equations (3.8) and (3.10) will only be satisfied if the arrival rate of class c customers is independent of the state of the queue. Thus quasi-reversibility differs from reversibility in that a stronger condition (3.8) is imposed on the arrival rates and a weaker condition (3.11) is imposed on the probability fluxes.

In the remainder of this section we shall extend our previous results on open networks of queues to apply to the case where the queues are quasi-reversible. If the network is of a certain fairly simple form this can be done easily. Suppose that customers pass through the network in accordance with routes determined by their types as described in the previous section. Associate with each customer arriving at queue j its class (i, s), i.e. its type and the stage of its route it has reached. Thus $j = r(i, s)$. Note that a customer's class does not alter while it passes through a queue, but changes as it moves from one queue to another. If the routes through the system allow the queues to be ordered so that a customer leaving a queue always moves to a queue later in the order (as in the case in Figs. 3.1 and 3.2) then the assumption of quasi-reversibility together with the arguments of Section 2.2 show that in equilibrium the states of the queues are independent. In this simple case the arrival streams at each queue are Poisson; we cannot hope for this to be true in more general networks, and so for these a different approach is required.

Let $\pi_j(\mathbf{x}_j)$ be the equilibrium distribution of a quasi-reversible queue at which arrivals of customers of class (i, s) form a Poisson process of rate $\alpha_j(i, s)$. Let $q_j(\mathbf{x}_j, \mathbf{x}_j')$ be the transition rates of this process and let $S_j(i, s, \mathbf{x}_j)$ be the set of states in which the queue contains one more customer of class (i, s) than in state $\mathbf{x}_j$, with the same number of customers of other classes. Consider now a Markov process $\mathbf{X}(t) = (\mathbf{x}_1(t), \mathbf{x}_2(t), \ldots, \mathbf{x}_J(t))$ whose transition rates are defined as follows. The probability intensity that a customer of type i enters the system and causes queue $k = r(i, 1)$ to change from state $\mathbf{x}_k$ to state $\mathbf{x}_k' \in \mathcal{S}_k(i, 1, \mathbf{x}_k)$ is $q_k(\mathbf{x}_k, \mathbf{x}_k')$. The probability intensity that a customer of type i leaves the system and causes queue $j = r(i, S(i))$ to change from state $\mathbf{x}_j' \in \mathcal{S}_j(i, S(i), \mathbf{x}_j)$ to state $\mathbf{x}_j$ is $q_j(\mathbf{x}_j', \mathbf{x}_j)$. The probability intensity that a customer of class (i, s), $s < S(i)$, leaves queue $j = r(i, s)$ and enters queue $k = r(i, s+1)$ as a customer of class $(i, s+1)$, causing queue j to change from state $\mathbf{x}_j' \in \mathcal{S}_j(i, s, \mathbf{x}_j)$ to state $\mathbf{x}_j$ and queue k to change from state $\mathbf{x}_k$ to state $\mathbf{x}_k' \in \mathcal{S}_k(i, s+1, \mathbf{x}_k)$, is

$$q_j(\mathbf{x}_j', \mathbf{x}_j) \frac{q_k(\mathbf{x}_k, \mathbf{x}_k')}{\sum_{\mathbf{x}' \in \mathcal{S}_k(i, s+1, \mathbf{x}_k)} q_k(\mathbf{x}_k, \mathbf{x}')} = q_j(\mathbf{x}_j', \mathbf{x}_j) \frac{q_k(\mathbf{x}_k, \mathbf{x}_k')}{\alpha_k(i, s+1)}$$

using equation (3.8). Finally, the probability intensity that there is an internal change in queue j from state $\mathbf{x}_j$ to state $\mathbf{x}_j'$, without the arrival or departure of any customer, is $q_j(\mathbf{x}_j, \mathbf{x}_j')$. The transition rates are thus defined in the obvious way: a queue behaves just as it would in isolation except that arrivals of class (i, s) customers, for $s > 1$, are triggered by departures from another queue rather than by an independent Poisson process. If queues

$1, 2, \ldots, J$ would in isolation be quasi-reversible and if the process $\mathbf{X}$ is in equilibrium then we shall call $\mathbf{X}$ an open network of quasi-reversible queues. Note that the jth queue of the network will not in general satisfy the conditions required for it to be quasi-reversible, and indeed the jth component of $\mathbf{X}$, $\mathbf{x}_j$, will not in general be a Markov process. Nevertheless, we shall occasionally abuse terminology and call queue j quasi-reversible—it would be if it were in isolation.

What might the reversed process $\mathbf{X}(-t)$ look like? The obvious possibility is that customers of type i might enter the system in a Poisson stream and pass backwards along their route and that the jth component of the system, $\mathbf{x}_j(-t)$, might be derived from the reversed version of queue j considered in isolation. Using Theorem 1.13 it becomes a routine matter to establish that this is indeed the reversed process and that the equilibrium distribution is

$$\pi(\mathbf{x}_1, \mathbf{x}_2, \ldots, \mathbf{x}_J) = \pi_1(\mathbf{x}_1)\pi_2(\mathbf{x}_2) \cdots \pi_J(\mathbf{x}_J) \tag{3.12}$$

The suggested probability intensity for the reversed process that a customer of class $(i, s+1)$ leaves queue $k = r(i, s+1)$ and enters queue $j = r(i, s)$ as a customer of class (i, s), causing queue k to change from state $\mathbf{x}'_k \in \mathcal{S}_k(i, s+1, \mathbf{x}_k)$ to state $\mathbf{x}_k$ and queue j to change from state $\mathbf{x}_j$ to state $\mathbf{x}'_j \in \mathcal{S}_j(i, s, \mathbf{x}_j)$, is

$$q'_k(\mathbf{x}'_k, \mathbf{x}_k)\frac{q'_j(\mathbf{x}_j, \mathbf{x}'_j)}{\sum_{\mathbf{x}' \in \mathcal{S}_j(i, s, \mathbf{x}_j)} q'_j(\mathbf{x}_j, \mathbf{x}')} = q'_k(\mathbf{x}'_k, \mathbf{x}_k)\frac{q'_j(\mathbf{x}_j, \mathbf{x}'_j)}{\alpha_j(i, s)}$$

from equation (3.10), which in turn followed from the quasi-reversibility of queue j. To establish condition (1.28) of Theorem 1.13 for transitions arising from the movement of customers from one queue to another we need therefore to show that

$$\frac{\pi_j(\mathbf{x}'_j)\pi_k(\mathbf{x}_k)q_j(\mathbf{x}'_j, \mathbf{x}_j)q_k(\mathbf{x}_k, \mathbf{x}'_k)}{\alpha_k(i, s+1)} = \frac{\pi_j(\mathbf{x}_j)\pi_k(\mathbf{x}'_k)q'_k(\mathbf{x}'_k, \mathbf{x}_k)q'_j(\mathbf{x}_j, \mathbf{x}'_j)}{\alpha_j(i, s)} \tag{3.13}$$

But this follows from equation (3.9) and the observation that $\alpha_k(i, s+1) = \alpha_j(i, s) = \nu(i)$. Condition (1.28) is established even more easily for transitions associated with the arrival at or departure from the system of a customer. The only remaining transitions are those where a single queue changes its state without the arrival or departure of a customer. Equation (3.9) establishes condition (1.28) directly for such transitions. We must finally check condition (1.27) of Theorem 1.13:

$$\begin{aligned} q(\mathbf{x}_1, \mathbf{x}_2, \ldots, \mathbf{x}_J) &= \sum_{j=1}^{J}\left(q_i(\mathbf{x}_j) - \sum_{(i,s)} \alpha_j(i,s)\right) + \sum_{i=1}^{I} \nu(i) \\ &= \sum_{j=1}^{J}\left(q'_j(\mathbf{x}_j) - \sum_{(i,s)} \alpha_j(i,s)\right) + \sum_{i=1}^{I} \nu(i) \\ &= q'(\mathbf{x}_1, \mathbf{x}_2, \ldots, \mathbf{x}_J) \end{aligned}$$

Thus the reversed process does take the conjectured form and the equilibrium distribution is given by expression (3.12).

As usual the form of the reversed process allows much to be deduced about the original process. The probability flux that a customer of class (i, s) departs from queue $j = r(i, s)$ and that queue j is left in state $\mathbf{x}_j$ is

$$\sum_{\mathbf{x}' \in \mathscr{S}_j(s, i, \mathbf{x}_j)} \pi_j(\mathbf{x}')q_j(\mathbf{x}'_j, \mathbf{x}_j) = \pi_j(\mathbf{x}_j)\alpha_j(i, s)$$

from equations (3.9) and (3.10). Thus if a customer of class (i, s) has just left queue j the probability he has left queue j in state $\mathbf{x}_j$ is $\pi_j(\mathbf{x}_j)$. The corresponding statement also holds for the reversed process, and hence a customer of class (i, s) arriving at queue $j = r(i, s)$ finds the queue in state $\mathbf{x}_j$ with probability $\pi_j(\mathbf{x}_j)$.

We can summarize the results of this section as follows.

Theorem 3.7. *An open network of quasi-reversible queues has the following properties:*

(i) *The states of the individual queues are independent.*

(ii) *For an individual queue the equilibrium distribution and the distribution over states found by an arriving customer of a given class are identical and are both as they would be if the queue were in isolation with arrivals of customers of each class forming independent Poisson processes.*

(iii) *Under time reversal the system becomes another open network of quasi-reversible queues.*

(iv) *The system itself is quasi-reversible and so departures from the system of customers of each type form independent Poisson processes, and the state of the system at time t_0 is independent of departures from the system prior to time t_0.*

Exercises 3.2

1. In the description of a quasi-reversible queue it was assumed that every customer who entered the queue left it, that customers did not change class as they passed through the queue, and that the process $\mathbf{x}(t)$ recorded how many customers of each class the queue contained. While these assumptions help us to visualize the queue they are not necessary. Show that the analysis of this section is unaltered if they are replaced by the weaker assumptions that the arrivals and departures of class c customers are signalled by transitions of the process $\mathbf{x}(t)$ and that the equilibrium arrival and departure rates of class c customers are equal, for $c \in \mathscr{C}$.
2. The definition of quasi-reversibility characterizes the Markov process representing the state of the queue, rather than any more fundamental property of the queue itself. It is quite possible that there may be two

representations of the same physical mechanism, one of which is quasi-reversible and the other not. Consider, for example, an isolated queue of the form described in the last section. Let the state of the queue be $(\mathbf{c}, c)$ where c is the class of the last customer to leave the queue. Show that with this representation the queue is not quasi-reversible.

3. Consider a stationary $M/M/1$ queue. Suppose that when a customer arrives at the queue a clerk issues him with a ticket and that when the customer leaves the queue he returns the ticket to the clerk (the purpose of the tickets may be to maintain the queue discipline). Now regard the clerk's office as a system in its own right and regard the tickets entering and leaving the office as customers. Show that although the arrival and departure streams are Poisson processes the system is not quasi-reversible however its state is defined, even under the weaker assumptions of Exercise 3.2.1.
4. Consider a queue with a Poisson arrival process and a state which is a reversible Markov process independent of future arrivals, e.g. the two-server queue considered in Section 1.5. Suppose now that each customer arriving at the queue is randomly allocated a class from the set $\mathscr{C}$, so that arrival times of class c customers, for $c \in \mathscr{C}$, form independent Poisson processes. Suppose further that the passage of a customer through the queue is unaffected by his class. Show that if the state of the queue is now taken to be the original reversible Markov process together with the classes of the customers in the queue arranged in order of their arrival, then the queue is quasi-reversible. If the passage of a customer through the queue *is* affected by his class then the queue may not be quasi-reversible however its state is defined, as the next exercise shows.
5. Arrivals of customers of types 1 and 2 at a single-server queue form independent Poisson processes. The service requirements of customers are independent and all have the same exponential distribution. The server gives priority to customers of type 1, and will even interrupt the service of a type 2 customer if a type 1 customer arrives. Deduce that departures from the queue form a Poisson process and that departures of type 1 customers form a Poisson process. Show that departures of type 2 customers do not form a Poisson process.
6. It was assumed early in this section that a customer entering a queue could not leave it immediately. Certain systems, e.g. the telephone exchange model of Section 2.1 or the queue with balking considered in Exercise 2.1.1, satisfy all the conditions for quasi-reversibility apart from this assumption. The assumption can be relaxed provided we deal with two technical difficulties. The first is that we must require that all arrivals and departures of class c customers are signalled by changes in the state of the queue, for each $c \in \mathscr{C}$. If $\mathscr{C}$ is finite it is easy to comply with this requirement using flip-flop variables as described in Section 2.1. The

second difficulty is that our definition of a network requires that a customer who enters a queue and leaves it immediately must go on to the next queue on his route. There will thus exist transitions of the Markov process $\mathbf{X}$ involving more than two queues. Extend the analysis of this section to deal with this difficulty. Observe that both difficulties can be avoided by using the method of Exercise 2.1.1(ii), whereby a customer pauses momentarily instead of leaving the queue immediately.

3.3 SYMMETRIC QUEUES

The quasi-reversible queues considered in Section 3.1 possess the property that the service requirement of a customer is exponentially distributed. This property simplifies analysis, since it removes the need for the state of the queue to include information on the amount of service customers have received. A more general distribution which can be handled with a little more effort is the gamma distribution. This arises when a customer requires a number of stages of service, each of which consists of an independent exponentially distributed amount of service. In this section we shall consider a range of queues which turn out to be quasi-reversible even when service requirements are not exponentially distributed. Initially we shall allow only service requirements which have a gamma distribution, but later we shall remove this restriction.

Consider a queue within which customers are ordered, with the queue containing customers in positions $1, 2, \ldots, n$, where n is the total number of customers in the queue. We shall call such a queue *symmetric* if it operates in the following manner:

(i) The service requirement of a customer is a random variable whose distribution may depend upon the class of the customer.

(ii) A total service effort is supplied at the rate $\phi(n)$.

(iii) A proportion $\gamma(l, n)$ of this effort is directed to the customer in position l $(l = 1, 2, \ldots, n)$; when this customer leaves the queue customers in positions $l+1, l+2, \ldots, n$ move to positions $l, l+1, \ldots, n-1$ respectively.

(iv) When a customer arrives at the queue he moves into position l $(l = 1, 2, \ldots, n+1)$ with probability $\gamma(l, n+1)$; customers previously in positions $l, l+1, \ldots, n$ move to positions $l+1, l+2, \ldots, n+1$ respectively.

Of course

$$\sum_{l=1}^{n} \gamma(l, n) = 1$$

and we shall insist that $\phi(n) > 0$ if $n > 0$. The queue described differs from those of Section 3.1 in that service requirements are not restricted and the symmetry condition $\gamma \equiv \delta$ is imposed. This condition rules out many queue disciplines, e.g. first come first served, and indeed at a symmetric queue there will be little queueing at all, in the usual sense of the word. Nevertheless, some useful systems can be set up as symmetric queues, and we shall describe four examples.

A server-sharing queue. When

$$\gamma(l, n) = \frac{1}{n} \qquad l = 1, 2, \ldots, n; \; n = 1, 2, \ldots$$

the service effort is shared equally between all customers in the queue. If $\phi(n) = 1$ for $n > 0$ then the queue behaves as a single-server queue, and a customer's remaining service requirement decreases at rate $1/n$.

A stack. When

$$\gamma(l, n) = 1 \qquad l = n; \; n = 1, 2, \ldots$$

the total service effort is directed to the customer who last arrived. Such a queue is best visualized as a stack, with customers arriving at and departing from the top of the stack. If $\phi(n) = 1$ for $n > 0$ then we have a single-server queue at which the queue discipline is last come first served with preemption (cf. Exercise 1.3.8).

A queue with no waiting room. Consider the functions

$$\phi(n) = n \qquad n = 1, 2, \ldots, K$$

$$\phi(n) = \xi \qquad n = K+1, K+2, \ldots$$

$$\gamma(l, n) = \frac{1}{n} \qquad l = 1, 2, \ldots, n; \; n = 1, 2, \ldots, K$$

$$\gamma(l, n) = 1 \qquad l = n; \; n = K+1, K+2, \ldots$$

where ξ is very large. We can regard this queue as one with K available servers at which a customer who arrives to find all K servers occupied leaves almost immediately. We have chosen not to make ξ infinite since this would entail a minor technical difficulty. It would allow an arrival and a departure to occur at the same time and not cause a change of state. This difficulty could be overcome using the flip-flop variable described in Section 2.1 in connection with the telephone exchange model, which would then be a special case.

An infinite-server queue. If

$$\phi(n) = n \qquad n = 1, 2, \ldots$$

$$\gamma(l, n) = \frac{1}{n} \qquad l = 1, 2, \ldots, n; \; n = 1, 2, \ldots$$

then the queue behaves as a queue with an infinite number of servers, with each customer having a server to himself; in this case customers do not affect each other within the queue. An infinite-server queue can be regarded as a special case of either a server-sharing queue or a queue with no waiting room.

Consider now a symmetric queue at which customers of class c arrive in a Poisson stream at rate $\nu(c)$. Suppose that a class c customer requires $w(c)$ stages of service, each of which consists of an independent exponentially distributed amount of service with mean $d(c)$. The service requirement of a class c customer will then have a gamma distribution, with mean $w(c)d(c)$ and variance $w(c)d(c)^2$.

Let $c(l)$ be the class of the customer in position l and suppose that his service has reached stage $u(l)$, where $1 \le u(l) \le w(c(l))$. Let $\mathbf{c}(l) = (c(l), u(l))$. Then

$$\mathbf{c} = (\mathbf{c}(1), \mathbf{c}(2), \ldots, \mathbf{c}(n))$$

(where n is the number in the queue) is a Markov process representing the state of the queue. We will now show that its equilibrium distribution is

$$\pi(\mathbf{c}) = b \prod_{l=1}^{n} \frac{\nu(c(l))d(c(l))}{\phi(l)} \tag{3.14}$$

provided the normalizing constant given by

$$b^{-1} = \sum_{n=0}^{\infty} \frac{a^n}{\prod_{l=1}^{n} \phi(l)} \tag{3.15}$$

where

$$a = \sum_{c} \nu(c)d(c)w(c)$$

is positive. Note that a is the average amount of service requirement arriving at the queue per unit time. It is fairly easy to show that expression (3.14) is the equilibrium distribution, since there is an obvious candidate for the reversed process, namely a queue at which arrivals are Poisson and which operates in precisely the same manner but with $u(l)$ recording the number of stages *yet to be completed* before the customer in position l leaves the queue. We shall now verify this. The probability intensity that a customer of class c arrives at the original queue and moves into position l is $\nu(c)\gamma(l, n+1)$, where n was the number previously in the queue. Let this event cause a transition from the state $\mathbf{c}$ to the state $\mathbf{c}'$. The probability intensity that when the state of the reversed queue is $\mathbf{c}'$ the customer in position l departs from the queue is $\phi(n+1)\gamma(l, n+1)/d(c)$. From the form

(3.14) we see that

$$\pi(\mathbf{c}') = \pi(\mathbf{c}) \frac{\nu(c)d(c)}{\phi(n+1)}$$

and hence that

$$\pi(\mathbf{c})\nu(c)\gamma(l, n+1) = \frac{\pi(\mathbf{c}')\phi(n+1)\gamma(l, n+1)}{d(c)}$$

Hence we can show that condition (1.28) of Theorem 1.13 holds for transitions arising from arrivals at the queue. Similarly, we can show that it holds for transitions caused by departures from the queue. The only remaining transitions are those which occur when an intermediate stage of a customer's service is completed. But if this causes a transition from $\mathbf{c}$ to $\mathbf{c}'$ then the transition rates $q(\mathbf{c}, \mathbf{c}')$ in the original process and $q(\mathbf{c}', \mathbf{c})$ in the reversed process are equal, and so are $\pi(\mathbf{c})$ and $\pi(\mathbf{c}')$. (Observe that it is the possibility of such a transition which differentiates the queue from those considered in Section 3.1 and which necessitates the symmetry condition $\gamma \equiv \delta$.) Finally, it is clear that $q(\mathbf{c}) = q'(\mathbf{c})$, and hence Theorem 1.13 shows that expression (3.14) does indeed give the equilibrium distribution and that the reversed process is of the suggested form. This in turn establishes that a symmetric queue in equilibrium is quasi-reversible, at least when service requirements have gamma distributions. In fact the queue is dynamically reversible with the conjugacy relation defined by

$$u^+(l) = w(c(l)) - u(l) + 1$$
$$\mathbf{c}^+(l) = (c(l), u^+(l))$$
$$\mathbf{c}^+ = (\mathbf{c}^+(1), \mathbf{c}^+(2), \ldots, \mathbf{c}^+(n))$$

If the sum in equation (3.15) is infinite then the queue cannot reach equilibrium: work arrives at the queue more quickly than it can be dealt with.

The equilibrium distribution (3.14) has some interesting implications. The probability there are n customers in the queue is

$$\frac{ba^n}{\prod_{l=1}^{n} \phi(l)} \tag{3.16}$$

Further, given there are n customers in the queue, $\mathbf{c}(1), \mathbf{c}(2), \ldots, \mathbf{c}(n)$ are independent. The customer in position l is of class c with probability

$$\frac{\nu(c)d(c)w(c)}{a} \tag{3.17}$$

and $u(l)$ is equally likely to be any value in the range $1 \leq u(l) \leq w(c(l))$. The constant a and the probabilities (3.16) and (3.17) depend on the values $d(c)$

and $w(c)$ only through the product $d(c)w(c)$, which is the mean of the service requirement distribution.

Suppose now that when a customer of class c arrives at the queue he is allocated a finer classification, (c, z), with probability $p(c, z)$, where z belongs to a countable set $\mathscr{Z}$, and $\sum_z p(c, z) = 1$ for each c. Then arrivals at the queue of customers of class (c, z) form a Poisson process of rate $\nu(c)p(c, z)$. If the service requirement of a customer of class (c, z) has a gamma distribution with mean $w(c, z)d(c, z)$ and variance $w(c, z)d(c, z)^2$, then the preceding analysis still applies with regard to the finer classification. The service requirement distribution of a customer of class c is now a gamma distribution with mean $w(c, z)d(c, z)$ and variance $w(c, z)d(c, z)^2$ with probability $p(c, z)$ for $z \in \mathscr{Z}$, i.e. it is a *mixture* of gamma distributions. The mean service requirement of a customer of class c is

$$a(c) = \sum_z p(c, z)w(c, z)d(c, z)$$

and the average amount of service requirement arriving at the queue per unit time is

$$\begin{aligned} a &= \sum_c \sum_z \nu(c)p(c, z)w(c, z)d(c, z) \\ &= \sum_c \nu(c)a(c) \end{aligned}$$

Let $\mathbf{c}(l) = (c(l), z(l), u(l))$ where $(c(l), z(l))$ is the refined classification of the customer in position l, and again take $\mathbf{c} = (\mathbf{c}(1), \mathbf{c}(2), \ldots, \mathbf{c}(n))$ to be the state of the queue. Then the equilibrium distribution is now

$$\pi(\mathbf{c}) = b \prod_{l=1}^{n} \frac{\nu(c(l))p(c(l), z(l))d(c(l), z(l))}{\phi(l)} \tag{3.18}$$

where b is defined as before by equation (3.15). Thus the probability there are n customers in the queue is again (3.16), and if there is a customer in position l the probability he is of class c is $\nu(c)a(c)/a$. These probabilities depend on the parameters $p(c, z)$, $d(c, z)$, and $w(c, z)$ defining the distribution of service requirement for a class c customer only through the mean service requirement $a(c)$. If we are given the class c of the customer in position l then the probability his refined classification is (c, z) is $p(c, z)d(c, z)w(c, z)/a(c)$. If we are given the refined classification (c, z) of the customer in position l then we will know $w(c, z)$, i.e. how many stages his service consists of; the number of the stage he has reached, $u(l)$, is equally likely to be any number between 1 and $w(c, z)$. Let us record some of these conclusions in the following theorem.

Theorem 3.8. *A stationary symmetric queue* $\mathbf{c}$ *at which service requirement distributions are mixtures of gamma distributions has the following* *properties:*

(i) *The probability the queue contains n customers is*

$$\frac{ba^n}{\prod_{l=1}^n \phi(l)}$$

(ii) *Given there are n customers in the queue the classes of the customers are independent and the probability the customer in a given position is of class c is*

$$\frac{\nu(c)a(c)}{a}$$

(iii) *The queue is quasi-reversible with respect to either the classification c or the refined classification* (c, z).

Suppose now that $\mathscr{Z}$ is a collection of positive numbers. For each $z \in \mathscr{Z}$ let $d(c, z)$ become very small and $w(c, z)$ very large, with $w(c, z)d(c, z)$ fixed at the value z. The variance of the gamma distribution associated with the refined classification (c, z), $w(c, z)d(c, z)^2$, tends to zero, while the mean remains at z. We can thus approximate a service requirement of exactly z. By using several refined classifications (c, z) for the class c it is possible to approximate as closely as we please an arbitrary distribution of service requirement; this is stated more precisely in the next result, which we shall prove in Exercise 3.3.3.

Lemma 3.9. *Let $F(x)$ be the distribution function of a positive random variable. Then it is possible to choose a sequence of distribution functions $F_m(x)$, each of which corresponds to a mixture of gamma distributions, so that*

$$\lim_{m \to \infty} F_m(x) = F(x)$$

for all x at which F is continuous.

Lemma 3.9 strongly suggests that Theorem 3.8 will remain valid without the restriction that service requirement distributions be mixtures of gamma distributions. This is in fact the case although we will not be able to prove it here, since a symmetric queue with arbitrary service requirement distributions cannot be represented by a Markov process with a countable state space. A continuous state space is required, and we shall have to content ourselves with a brief sketch of the results.

Consider then a symmetric queue at which customers of class c arrive in a Poisson stream of rate $\nu(c)$, and suppose the service requirement distribution of a class c customer has distribution function $F_c(x)$, with mean $a(c)$. Let

$$a = \sum_c \nu(c)a(c)$$

and define b as before by equation (3.15). Let $c(l) \in \mathscr{C}$ be the class of the customer in position l, let $z(l) \in (0, \infty)$ be his service requirement, and let $u(l) \in (0, z(l))$ be the amount of service effort he has so far received. Let $\mathbf{c}(l) = (c(l), z(l), u(l))$ and take $\mathbf{c} = (\mathbf{c}(1), \mathbf{c}(2), \ldots, \mathbf{c}(n))$ to be the state of the queue. Observe that the process $\mathbf{c}$ is Markov with a continuous state space. Jumps in the process $\mathbf{c}$ occur when a customer arrives or departs, but between these jumps $\mathbf{c}$ changes continuously, with $u(l)$ increasing linearly at rate $\phi(n)\gamma(l, n)$. When $u(l)$ reaches $z(l)$ the customer in position l leaves the queue.

Theorem 3.10. *A stationary symmetric queue* $\mathbf{c}$ *at which service requirements are arbitrarily distributed has properties* (i) *to* (iii) *listed in Theorem 3.8. In addition:*

(iv) *Given the number of customers in the queue and the class of each of them, the amounts of service effort the customers have received are independent, and the probability a customer of class* c *has received an amount of service effort not greater than* x *is*

$$F_c^*(x) = \frac{1}{a(c)} \int_0^x (1 - F_c(z))\, dz$$

Outline of proof. The theorem is proved by showing that the equilibrium distribution is the probability density

$$b \prod_{l=1}^{n} \frac{du(l)\, dF_{c(l)}(z(l))}{\phi(l)} \nu(c(l)) \tag{3.19}$$

$n = 0, 1, 2, \ldots, c(l) \in \mathscr{C}, 0 < z(l) < \infty, 0 < u(l) < z(l), l = 1, 2, \ldots, n$, and that the reversed process is an identical symmetric queue but with $u(l)$ recording the amount of service effort yet to be received by the customer in position l. These facts can be established by a direct consideration of the process $\mathbf{c}$ or by a limiting argument based on a sequence of symmetric queues chosen so that the limit of the sequence is the process $\mathbf{c}$, but where at each queue in the sequence service requirement distributions are mixtures of gamma distributions.

The equilibrium distribution (3.19) is a density with respect to $u(1), u(2), \ldots, u(n), z(1), z(2), \ldots, z(n)$ for each value of n and for each arrangement $(c(1), c(2), \ldots, c(n))$. Properties (i), (ii), and (iv) follow by integrating this density over the appropriate values of these variables. For example the distribution (3.19) shows that, given the number of customers in the queue and the class of each of them, $z(l)$ is distributed with density

$$\frac{1}{a(c(l))} z(l)\, dF_{c(l)}(z(l)) \qquad 0 < z(l) < \infty$$

and, given $z(l)$, $u(l)$ is distributed uniformly on $(0, z(l))$. Hence the probability a customer of class c has received an amount of service effort greater than x is

$$\frac{1}{a(c)}\int_x^\infty\left[\int_x^z du\right]dF_c(z)$$

which reduces to $1-F_c^*(x)$.

The probabilities in (i) and (ii) are *insensitive* to the form of the distribution functions $F_c(x)$ in that they depend upon them only through their means $a(c)$. Thus, for example, Erlang's formula (1.13), calculated for a telephone exchange model in which call lengths are exponentially distributed, holds even when call lengths are arbitrarily distributed since the model is a symmetric queue.

The distribution function $F_c^*(x)$ is familiar as the equilibrium age distribution of a renewal process in which components have lifetime distribution $F_c(x)$. We can in fact derive this from our results. Consider a queue with no waiting room and with just one server. Suppose that there is just one class of customer c and that the arrival rate $\nu(c)$ is very large. This queue is equivalent to a renewal process, since as soon as the single customer in the queue is served (a component fails) he is replaced by another customer (a new component). Thus in equilibrium the age of the component in use has distribution function $F_c^*(x)$.

The above example illustrates a minor difficulty which can arise with arbitrary distributions. If lifetimes are all the same fixed constant, the renewal process is periodic and will not approach equilibrium unless it starts there. Periodicity cannot arise when distributions are mixtures of gamma distributions, but must be watched for in general.

The number in a symmetric queue will not usually be a Markov process. Nevertheless, the form of the reversed process leads immediately to the following result.

Theorem 3.11. *The number in a stationary symmetric queue is a reversible stochastic process.*

This property, like quasi-reversibility, is lost when the queue is part of a network of queues.

A major difference between the symmetric queues considered in this section and the queues considered in Section 3.1 is that at a symmetric queue the service requirement of a customer can depend upon his class. In a network of quasi-reversible queues this has two important consequences which we shall explore further in the next chapter. First, a customer's service requirement at a symmetric queue can depend upon the queues he has previously visited and the queues he has yet to visit. This follows naturally since his class depends upon his type, which determines his route through the queues of the network. Hence if the type of a customer in a symmetric queue is unknown then his future route can depend upon his service requirement at that queue. Second, given a customer's route through the queues of the network, his service requirements at symmetric queues along that route may be dependent. This will happen when a variety of customer types correspond to the same route through the system but to different service requirements at the symmetric queues along that route. Indeed, by using enough customer types it is possible to approximate as closely as we please any desired pattern of dependence between the service requirements at the symmetric queues along a route (Exercise 3.3.12) and between these service requirements and the route itself. This strongly suggests that arbitrary patterns of dependence can be allowed, but once again to establish this would take us beyond the realm of countable state space Markov processes.

Although symmetric queues are not the only quasi-reversible queues whose operation involves arbitrary distributions (see, for example, Exercise 3.5.11), they form a class large enough to include all the special cases of such queues which we shall need in the next chapter when we discuss examples of queueing networks.

Exercises 3.3

1. The queue with no waiting room described above has the property that when a customer arrives to find all the servers occupied the servers pause momentarily until he leaves. Redefine the functions ϕ and γ so that this does not occur.
2. Show how the functions ϕ and γ can be chosen to represent a server-sharing queue with a maximum size of N, so that customers who arrive when N customers are being served are turned away.
3. Prove Lemma 3.9, using

$$F_m(x)=\sum_{k=1}^{\infty}\left[F\left(\frac{k}{m}\right)-F\left(\frac{k-1}{m}\right)\right]G_m^k(x) \qquad x\geq 0$$

where

(i) G_m^k is the distribution function of a gamma distribution with mean k/m and variance k/m^2 (i.e. the distribution function of the sum of k exponential random variables each with mean $1/m$),

(ii) G_m^k is the distribution function of a gamma distribution with mean $g(k, m)$ and variance $g(k, m)/m$ where

$$g(k, m)=\int_{(k-1)/m}^{k/m}\frac{x\,dF(x)}{F(k/m)-F((k-1)/m)}$$

Show that in this case the mean of F_m is equal to the mean of F.

4. Show that $F(x) = F^*(x)$ if and only if F is the exponential distribution.
5. If $F_c(x)$, $c \in \mathscr{C}$, correspond to mixtures of gamma distributions establish property (iv) of Theorem 3.10 directly from the equilibrium distribution (3.18).
6. Show that the service requirement yet to be received by a customer in a symmetric queue has mean

$$\frac{\mu^2+\sigma^2}{2\mu}$$

where μ and σ^2 are the mean and variance respectively of the service requirement distribution.
7. Suppose that an $M/G/1$ queue has a queue discipline which allows it to be considered as a symmetric queue, e.g. last come first served with preemption or server sharing. Let W be the service requirement yet to be received summed over all the customers in the queue. Show that the distribution of W is the same as that of a geometric sum of independent random variables each with distribution function F^*. Use the previous exercise to deduce that

$$E(W) = \frac{\nu(\mu^2+\sigma^2)}{2(1-\nu\mu)} \tag{3.20}$$

where ν is the arrival rate at the queue. Observe that the distribution of W does not depend upon the queue discipline. If the queue discipline is first come first served W is called the virtual waiting time and is the time a typical customer would have to queue to before his service started; expression (3.20) is known as the Pollaczek–Khinchin formula.
8. Show that the examples of symmetric queues given in this section can each be constructed from the queues of Section 3.1. Begin by showing that a service requirement with a gamma distribution can be obtained by requiring a customer to pass through the same queue a fixed number of times before moving on to the next queue.
9. Show that the waiting time of a customer at a stack has the same distribution as the busy period in an $M/G/1$ queue.
10. Show that at either the server-sharing queue or the queue with no waiting room the order of the customers in the queue is independent of the order of arrival of these customers. Suppose now that customers of class c require an amount of service which is exponentially distributed with mean $1/\lambda(c)$. If the queue contains n customers show that the probability the customers arrived in a given order is

$$\prod_{l=1}^{n} \frac{\lambda(c(l))}{\sum_{m=1}^{l} \lambda(c(m))}$$

where $c(l)$, $l = 1, 2, \ldots, n$, is the class of the customer who is lth in the given order.

11. Suppose that in the telephone exchange model considered in Sections 1.3 and 2.1 call lengths are arbitrarily distributed. Show that the points in time at which a call is lost or is completed form a Poisson process. Show that the points in time at which a call is lost form a reversible point process. Establish the result stated in Exercise 2.1.4.
12. Let $F(x_1, x_2, \ldots, x_n)$ be the joint distribution function of n positive random variables. Show that if

$$F_m(x_1, x_2, \ldots, x_n) = \sum_{k_1=1}^{\infty} \sum_{k_2=1}^{\infty} \cdots \sum_{k_n=1}^{\infty} G_m^{k_1}(x_1) G_m^{k_2}(x_2) \cdots G_m^{k_n}(x_n) \times \int_{(k_1-1)/m}^{k_1/m} \int_{(k_2-1)/m}^{k_2/m} \cdots \int_{(k_n-1)/m}^{k_n/m} dF(z_1, z_2, \ldots, z_n)$$

where G_m^k is defined as in either part (i) or (ii) of Exercise 3.3.3 then

$$\lim_{m \to \infty} F_m(x_1, x_2, \ldots, x_n) = F(x_1, x_2, \ldots, x_n)$$

for all $(x_1, x_2, \ldots, x_n)$ at which F is continuous.
13. Consider a symmetric queue within an open network of quasi-reversible queues. Suppose that customers enter the system at rate ν and that the mean service requirement of a customer at the symmetric queue, summed over all the customer's visits to the queue and averaged over all customer types, is a. Show that the equilibrium distribution for the number in the symmetric queue is just what it would be if that queue were in isolation and customers with mean service requirement a arrived in a Poisson stream at rate ν.
14. Consider a symmetric queue within an open network of quasi-reversible queues. Deduce from Theorem 3.8(ii) that the mean number of class c customers in the symmetric queue is proportional to $\nu(c)a(c)$. By supposing that the classification c is fine enough to determine the customer's service requirement deduce from Theorem 3.10 and Little's result that the mean period a customer spends in the symmetric queue is proportional to his service requirement there.
15. Show that for a stack the function $\phi(n)$ can be replaced by a function $\phi(c(1), c(2), \ldots, c(n))$ without destroying the property of quasi-reversibility.

3.4 CLOSED NETWORKS

In the open networks considered previously in this chapter customers of type i entered the system in a Poisson stream at rate $\nu(i)$ and passed through the sequence of queues

$$r(i, 1), r(i, 2), \ldots, r(i, S(i))$$

before leaving the system. Suppose now that customers of type i return to queue $r(i, 1)$ after leaving queue $r(i, S(i))$ and repeat their route through the system. The network will become closed, with customers neither entering nor leaving the system. The number of customers of type i within the system, $N(i)$ say, will remain fixed for $i = 1, 2, \ldots, I$.

If the queues of the system would in isolation be quasi-reversible the equilibrium distribution takes a simple form. Let $\pi_j(\mathbf{x}_j)$ again be the equilibrium distribution of the jth queue in the network if it were in isolation, with arrivals of customers of class (i, s) forming a Poisson process of rate $\alpha_j(i, s)$. Let

$$\pi(\mathbf{x}_1, \mathbf{x}_2, \ldots, \mathbf{x}_J) = B\pi_1(\mathbf{x}_1)\pi_2(\mathbf{x}_2) \cdots \pi_J(\mathbf{x}_J) \tag{3.21}$$

where B is chosen so that the distribution π sums to unity, with the sum taken over $\mathcal{S}$, the set of states $(\mathbf{x}_1, \mathbf{x}_2, \ldots, \mathbf{x}_J)$ for which the total number of type i customers is $N(i)$, for $i = 1, 2, \ldots, I$. It is fairly easy to check that π is the equilibrium distribution for the system and that the reversed process consists of the reversed queues with customers moving backwards around their routes. Indeed we have done most of the work already in Section 3.2 when dealing with open networks. The relations established there for transitions arising from the movement of a customer from one queue to another (equation 3.13) or from internal changes in a queue (equation 3.9) apply here also. The movement of a type i customer from queue $r(i, S(i))$ to queue $r(i, 1)$ can be dealt with in precisely the same way as a movement from queue $r(i, s)$ to $r(i, s+1)$. Finally,

$$\begin{aligned} q(\mathbf{x}_1, \mathbf{x}_2, \ldots, \mathbf{x}_J) &= \sum_{j=1}^{J} \left(q_j(\mathbf{x}_j) - \sum_{(i,s)} \alpha_j(i,s) \right) \\ &= \sum_{j=1}^{J} \left(q_j'(\mathbf{x}_j) - \sum_{(i,s)} \alpha_j(i,s) \right) \\ &= q'(\mathbf{x}_1, \mathbf{x}_2, \ldots, \mathbf{x}_J) \end{aligned} \tag{3.22}$$

and so Theorem 1.13 establishes the desired result.

Some comments on the distribution (3.21) are in order. First, since $(\mathbf{x}_1, \mathbf{x}_2, \ldots, \mathbf{x}_J)$ is constrained to lie in the set $\mathcal{S}$ it does not follow from (3.21) that $\mathbf{x}_1, \mathbf{x}_2, \ldots, \mathbf{x}_J$ are independent. Second, the distributions $\pi_1(\mathbf{x}_1), \pi_2(\mathbf{x}_2), \ldots, \pi_J(\mathbf{x}_J)$ depend upon $\nu(i)$, $i = 1, 2, \ldots, I$, yet the distribution $\pi(\mathbf{x}_1, \mathbf{x}_2, \ldots, \mathbf{x}_J)$ cannot since in a closed network the values of these parameters do not affect the process. The resolution of this apparent contradiction lies in the role of the normalizing constant B. If $\nu(i)$ is changed then $\pi_1(\mathbf{x}_1), \pi_2(\mathbf{x}_2), \ldots, \pi_J(\mathbf{x}_J)$ do indeed change, but so does B, and B changes in such a way that $\pi(\mathbf{x}_1, \mathbf{x}_2, \ldots, \mathbf{x}_J)$ remains unaltered (cf. the discussion following Theorem 2.3).

In an open network we represented the behaviour of a customer with a random route by using a set of types, one of which was allocated to him at random (see Fig. 3.2). In the closed network just described a customer's type never changes, and this prevents us from modelling the network illustrated in Fig. 3.3 where a customer leaving queue 2 chooses at random whether to go to queue 1 or queue 3. We shall solve this problem by allowing a customer's type to change randomly in a closed network. Thus for the network in Fig. 3.3 suppose there are two customer types, and suppose a customer of type 1 follows the route 1, 2 and a customer of type 2 the route 3, 2. If we allow a customer leaving queue 2 to choose his type at random from the set $\{1, 2\}$ then customers will behave in the required way.

More generally, suppose the set of types $\{1, 2, \ldots, I\}$ is divided into disjoint subsets $\mathscr{I}(1), \mathscr{I}(2), \ldots,$ and that on leaving queue $r(i, S(i))$ a customer of type $i \in \mathscr{I}(m)$ becomes a customer of type $i' \in \mathscr{I}(m)$ with probability

$$\frac{\nu(i')}{\sum_{i \in \mathscr{I}(m)} \nu(i)} \tag{3.23}$$

He then proceeds through queues $r(i', 1), r(i', 2), \ldots, r(i', S(i'))$ before rechoosing his type again. His type will always belong to the set $\mathscr{I}(m)$ and thus

$$M(m) = \sum_{i \in \mathscr{I}(m)} N(i)$$

will remain constant for $m = 1, 2, \ldots$. The above mechanism will allow very general routing schemes and certainly those which can arise in closed migration processes (Exercise 3.4.1). It would be possible to regard m as the type of a customer and $i \in \mathscr{I}(m)$ as a finer indication of his progress: with this terminology the type of a customer would not keep changing. We prefer to call i the type since for open networks at least it is helpful to have the route of a customer determined by his type. Observe that in a closed network the points in time at which a customer rechooses his type can be regarded as regeneration points for him.

Consider then a system $\mathbf{X} = (\mathbf{x}_1, \mathbf{x}_2, \ldots, \mathbf{x}_J)$ with the above routing mechanism and containing queues which would in isolation be quasi-reversible. If the process $\mathbf{X} = (\mathbf{x}_1, \mathbf{x}_2, \ldots, \mathbf{x}_J)$ is in equilibrium then call $\mathbf{X}$ a closed network of quasi-reversible queues. An obvious possibility for the equilibrium distribution is

$$\pi(\mathbf{x}_1, \mathbf{x}_2, \ldots, \mathbf{x}_J) = B(M(1), M(2), \ldots)\pi_1(\mathbf{x}_1)\pi_2(\mathbf{x}_2) \cdots \pi_J(\mathbf{x}_J) \tag{3.24}$$

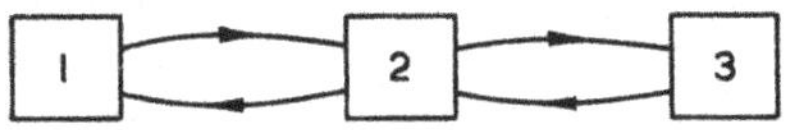

Fig. 3.3 A closed network

where the normalizing constant $B(M(1), M(2), \ldots)$ is chosen so that the distribution π sums to unity, with the sum taken over the state space

$$\mathscr{S}(M(1), M(2), \ldots) = \left\{(\mathbf{x}_1, \mathbf{x}_2, \ldots, \mathbf{x}_J) \,\middle|\, \sum_{i \in \mathscr{I}(m)} N(i) = M(m), \qquad m = 1, 2, \ldots\right\}$$

We shall use Theorem 1.13 to check that (3.24) is the equilibrium distribution and that the reversed process consists of the reversed queues with customers moving backwards around their routes and changing from type i to type i' after leaving queue $r(i, 1)$ with probability (3.23). Consider the transition which arises when a customer of type i leaves queue $j = r(i, S(i))$, changes to type i', and enters queue $k = r(i', 1)$. To establish condition (1.28) for this transition we must show that

$$\begin{aligned}\pi_j(\mathbf{x}_j')\pi_k(\mathbf{x}_k)q_j(\mathbf{x}_j', \mathbf{x}_j)&\frac{q_k(\mathbf{x}_k, \mathbf{x}_k')}{\alpha_k(i', 1)}\frac{\nu(i')}{\sum_{i \in \mathscr{I}(m)} \nu(i)}\\ &= \pi_j(\mathbf{x}_j)\pi_k(\mathbf{x}_k')q_k'(\mathbf{x}_k', \mathbf{x}_k)\frac{q_j'(\mathbf{x}_j, \mathbf{x}_j')}{\alpha_j(i, S(i))}\frac{\nu(i)}{\sum_{i \in \mathscr{I}(m)} \nu(i)}\end{aligned} \tag{3.25}$$

All but the final terms in equation (3.25) arise in the same way as did the corresponding terms in equation (3.13). But since $\alpha_k(i', 1) = \nu(i')$ and $\alpha_j(i, S(i)) = \nu(i)$, equation (3.25) reduces to equation (3.13) which has already been established. For all the other transitions arising in a closed network condition (1.28) takes the same form as for the corresponding transitions in an open network and has thus already been established. Finally, condition (1.27) follows from equation (3.22). Hence the equilibrium distribution and the reversed process are of the suggested form.

Consider now the instant at which a customer of class (i, s), $i \in \mathscr{I}(m)$, leaves queue $j = r(i, s)$. Let $\mathbf{x}_k$, for $k \neq j$, be the state of queue k immediately before this instant. Call $(\mathbf{x}_1, \mathbf{x}_2, \ldots, \mathbf{x}_J)$ the disposition of the other customers in the system; note that it is a member of the set $\mathscr{S}(M(1), M(2), \ldots, M(m)-1, \ldots)$, since one customer of type $i \in \mathscr{I}(m)$ is not included in the description $(\mathbf{x}_1, \mathbf{x}_2, \ldots, \mathbf{x}_J)$. The probability flux that a customer of class (i, s) departs from queue $j = r(i, s)$ with $(\mathbf{x}_1, \mathbf{x}_2, \ldots, \mathbf{x}_J)$ the disposition of the other customers in the system is

$$\begin{aligned}B(M(1), M(2), \ldots)\left(\prod_{k \neq j} \pi_k(\mathbf{x}_k)\right) &\sum_{\mathbf{x}' \in \mathscr{S}_j(s, i, \mathbf{x}_j)} \pi_j(\mathbf{x}')q_j(\mathbf{x}', \mathbf{x}_j)\\ &= B(M(1), M(2), \ldots)\pi_1(\mathbf{x}_1)\pi_2(\mathbf{x}_2) \cdots \pi_J(\mathbf{x}_J)\alpha_j(i, s)\end{aligned}$$

from equations (3.8) and (3.9). Thus if a customer of class (i, s) has just left queue j the probability that the disposition of the other customers in the system is $(\mathbf{x}_1, \mathbf{x}_2, \ldots, \mathbf{x}_J)$ is proportional to $\pi_1(\mathbf{x}_1)\pi_2(\mathbf{x}_2) \cdots \pi_J(\mathbf{x}_J)$. The constant of proportionality is found by summing over the set

$\mathcal{S}(M(1), M(2), \ldots, M(m)-1, \ldots)$ and is hence $B(M(1), M(2), \ldots, M(m)-1, \ldots)$. This is an intriguing result: the disposition of the other customers in the system is distributed in accordance with the equilibrium distribution which would obtain if they were the only customers in the system. Consideration of the reversed process shows that the same statement is valid at the instant when a customer of class (i, s) *arrives* at queue $j = r(i, s)$.

We can summarize the results of this section as follows.

Theorem 3.12. *A closed network of quasi-reversible queues has the following properties:*

(i) *The equilibrium distribution is of the form* (3.24).

(ii) *Under time reversal the system becomes another closed network of quasi-reversible queues.*

(iii) *When a customer of a given class arrives at a queue the disposition of the other customers in the system is distributed in accordance with the equilibrium distribution which would obtain if they were the only customers in the system.*

If a closed network of quasi-reversible queues contains symmetric queues then by using various customer types as described in Section 3.3 it is possible to allow dependences between a customer's service requirements at the symmetric queues he visits and between these service requirements and his route. We shall discuss this point further in Section 4.2, where it can be illustrated with some simple examples.

Exercises 3.4

1. Consider a closed migration process with transition rates (2.1) where $\lambda_j = 1$, $j = 1, 2, \ldots, J$. Show that the process can be regarded as a closed queueing network with customers whose route will be $r(i, 1), r(i, 2), \ldots, r(i, S(i))$, where $r(i, 1) = 1$, $r(i, s) \neq 1$ for $s \neq 1$, having the parameter

$$\nu(i) = \lambda_{r(i,1),r(i,2)}\lambda_{r(i,2),r(i,3)} \cdots \lambda_{r(i,S(i)-1),r(i,S(i))}\lambda_{r(i,S(i)),r(i,1)}$$

Show that the quantities $a_1, a_2, \ldots, a_J$ calculated from the queueing network parameters are proportional to the quantities $\alpha_1, \alpha_2, \ldots, \alpha_J$ determined by equations (2.2) from the migration process parameters.

2. Let

$$\pi_j(M_j(1), M_j(2), \ldots) = \sum \pi_j(\mathbf{x}_j)$$

where the summation runs over all $\mathbf{x}_j$ such that queue j contains $M_j(m)$ customers whose type is in the set $\mathcal{I}(m)$, for $m = 1, 2, \ldots$. Show that the equilibrium distribution (3.24) implies that the probability queue j

contains $M_j(m)$ customers of type $i \in \mathcal{I}(m)$, for $m = 1, 2, \ldots$, $j = 1, 2, \ldots, J$, is

$$B(M(1), M(2), \ldots) \prod_{j=1}^{J} \pi_j(M_j(1), M_j(2), \ldots)$$

provided

$$\sum_j M_j(m) = M(m) \qquad \text{for } m = 1, 2, \ldots$$

3. The normalizing constant $B(M(1), M(2), \ldots)$ can be calculated using the generating function method introduced in Exercise 2.3.6. Define the generating functions

$$\Phi_j(z(1), z(2), \ldots) = \sum_m \sum_{M_j(m)} \pi_j(M_j(1), M_j(2), \ldots) z(1)^{M_j(1)} z(2)^{M_j(2)} \cdots$$

$$B(z(1), z(2), \ldots) = \sum_m \sum_{M(m)} \frac{z(1)^{M(1)} z(2)^{M(2)} \cdots}{B(M(1), M(2), \ldots)}$$

Show that

$$B(z(1), z(2), \ldots) = \prod_j \Phi_j(z(1), z(2), \ldots)$$

4. Consider a closed network of quasi-reversible queues containing a symmetric queue. The service requirement of a customer visiting the symmetric queue may well be dependent on his earlier service requirements at this and other symmetric queues, because these may all be related to his type. Use Exercise 3.4.2 to show that the equilibrium distribution for the number of customers in the various queues of the network is insensitive to these dependencies and that it will be unaltered if each customer arriving at the symmetric queue has a service requirement independent of his previous service requirements so long as the mean service requirement of each customer at the queue is unaltered.
5. By considering the probability flux that a customer of class (i, s), $i \in \mathcal{I}(m)$, leaves queue $j = r(i, s)$ show that in equilibrium the mean arrival rate of customers of class (i, s) at queue j is

$$\frac{\alpha_j(i, s) B(M(1), M(2), \ldots, M(m), \ldots)}{B(M(1), M(2), \ldots, M(m)-1, \ldots)}$$

6. Exercise 3.3.14 has a parallel for a closed network. Consider a symmetric queue within a closed network of quasi-reversible queues. Show that for a given customer entering the symmetric queue the mean period he will remain in the queue is proportional to his service requirement there. Observe that, unlike the parallel result for open networks, the constant of proportionality depends on the chosen customer. This is not surprising: when a given customer arrives at a symmetric queue the

disposition of the other customers does not depend on the class of the given customer if the network is open, but it does if the network is closed.

7. A haulage firm has a fleet of N lorries, but at any one time some of these are being overhauled. While there are n lorries available the firm is able to handle an amount $\phi(n)$ of work, measured in miles per day, and this is shared out equally between the n lorries. A lorry requires an overhaul when it has travelled X miles since its last overhaul. An overhaul takes Y days. The quantities X and Y may be random variables; if so these random variables are independent between different lorries, but values relating to the same lorry may be dependent. Derive an expression for the mean amount of work handled per day by the firm.
8. A complex device consists of n main units, all of which must be operative for the normal operation of the device. Each main unit may fail within time h with probability $\lambda h + o(h)$, independently of the other service units and of its previous service life. There are $m+l$ additional units, m of which are active, i.e. may fail with the same probability as the main units, while the remaining l are passive and cannot fail. Failed units are sent for repair and take a mean time μ^{-1} to repair. If some of the main units fail these are replaced by units from the active redundant system, and these in turn by units of the passive redundant system. Describe the system as a closed network of queues and obtain the equilibrium probability that the device is operative.
9. Consider the model of a mining operation described in Section 2.3. Suppose that after machine j has dealt with face i the face must be left for a period until the dust has settled before the next machine can start work on it; let this period have mean X_{ij}. Obtain the equilibrium distribution for the system.
10. Consider again the model of mining operation described in Section 2.3. The model assumed that it took no time for a machine to travel from one face to the next. Suppose now that it takes a mean period Y_{ij} for machine j to travel from face i to the next face. Obtain the equilibrium distribution for the system when $\phi_1 = \phi_2 = \cdots = \phi_J = \phi$.
11. All the queues we have considered so far which convert a Poisson arrival process into a Poisson departure process have the property that the number in the queue is a reversible stochastic process. This property does not follow necessarily, as the following example shows. Consider a closed network with two customers and two queues. Queue 1 would in isolation behave as a first come first served $M/M/1$ queue. Queue 2 would in isolation behave as a queue with one server and no waiting room. The first customer's service requirement at queue 2 is 1. The second customer's service requirement at queue 2 changes in the cycle

$2, 3, \ldots, s, 2, 3, \ldots$. Both customers' service requirements at queue 1 are independent exponentially distributed random variables with mean λ^{-1}. Show that the arrival process at and the departure process from queue 2 are both Poisson. Show that the number of customers in queue 2 is not a reversible stochastic process. Observe that these statements are true for queue 1 as well.

3.5 MORE GENERAL ARRIVAL RATES

In the open networks considered in Section 3.2 the state of the system did not affect the streams of customers entering the system. In this section we shall show that results can be obtained when the rates of arrival at the system are influenced by the state of the system, provided the influence takes a certain fairly restricted form.

It is illuminating to approach these results by consideration of the following queue. Suppose a queue is such that its state is $(n(1), n(2), \ldots)$, where $n(c)$ is the number of customers of class c that it contains, and suppose arrivals of class c customers form a Poisson stream of rate $\nu(c)$, $c \in \mathscr{C}$, where the arrival streams are independent of each other. Write

$$\nu(c)\phi_c(n(1), n(2), \ldots)$$

for the probability intensity that a customer of class c leaves the queue when its state is $(n(1), n(2), \ldots)$.

Lemma 3.13. *If the above queue is in equilibrium the following statements are equivalent:*

(i) *The process* $(n(1), n(2), \ldots)$ *is reversible.*

(ii) *The queue is quasi-reversible.*

(iii) *There exists a function* $\Phi(n(1), n(2), \ldots)$ *such that*

$$\Phi(n(1), n(2), \ldots, n(c), \ldots) = \phi_c(n(1), n(2), \ldots, n(c), \ldots) \times \Phi(n(1), n(2), \ldots, n(c)-1, \ldots) \tag{3.26}$$

Proof. Let

$$\pi(n(1), n(2), \ldots) = \frac{b}{\Phi(n(1), n(2), \ldots)}$$

If relation (3.26) holds then π satisfies the detailed balance conditions; hence statement (iii) implies statement (i). Conversely, if the process is reversible the equilibrium distribution $\pi(n(1), n(2), \ldots)$ satisfies the detailed balance condition

$$\pi(n(1), n(2), \ldots, n(c)-1, \ldots)\nu(c) = \pi(n(1), n(2), \ldots, n(c), \ldots) \times \nu(c)\phi_c(n(1), n(2), \ldots, n(c), \ldots)$$

and hence

$$\Phi(n(1), n(2), \ldots) = \pi(n(1), n(2), \ldots)^{-1}$$

satisfies the relation (3.26). Thus statement (i) implies statement (iii).

Statement (i) immediately implies statement (ii). Conversely, if the queue is quasi-reversible then equation (3.11) shows that the detailed balance condition is satisfied, and thus statement (ii) implies statement (i).

Consider now a closed network in which customers do not change type, of the sort discussed early in the previous section. Suppose that a quasi-reversible queue of the above form is appended to the network as queue 0, and that a visit to this queue is added as an extra stage at the beginning of the route of each customer; thus $r(i, 0) = 0$, for $i = 1, 2, \ldots$. Let $N(i)$ be the number of customers of type i in queues $1, 2, \ldots, J$ and $N^*(i)$ the number in queues $0, 1, 2, \ldots, J$. Thus $N^*(i) = N(i) + n(i)$. Now define a function Ψ by

$$\Psi(N(1), N(2), \ldots) = \begin{cases} [\Phi(n(1), n(2), \ldots)]^{-1} & \text{if } N(i) \le N^*(i),\ i = 1, 2, \ldots \\ 0 & \text{otherwise} \end{cases}$$

where $n(i) = N^*(i) - N(i)$. If we confine our attention to queues $1, 2, \ldots, J$ how does the system behave? If these queues contain between them $N(i)$ customers of type i, for $i = 1, 2, \ldots$, then the probability intensity a customer of type i leaves queue 0 and enters the system is

$$\nu(i) \frac{\Phi(n(1), n(2), \ldots, n(i), \ldots)}{\Phi(n(1), n(2), \ldots, n(i)-1, \ldots)} = \nu(i) \frac{\Psi(N(1), N(2), \ldots, N(i)+1, \ldots)}{\Psi(N(1), N(2), \ldots, N(i), \ldots)} \tag{3.27}$$

The equilibrium distribution for the system is, by Theorem 3.12, of the form

$$B \frac{1}{\Phi(n(1), n(2), \ldots)} \pi_1(\mathbf{x}_1)\pi_2(\mathbf{x}_2) \cdots \pi_J(\mathbf{x}_J) = B\Psi(N(1), N(2), \ldots)\pi_1(\mathbf{x}_1)\pi_2(\mathbf{x}_2) \cdots \pi_J(\mathbf{x}_J) \tag{3.28}$$

The above discussion strongly suggests that in an open network of queues if we relax the assumption of Poisson arrival streams, and suppose instead that the probability intensity a customer of type i arrives takes the form (3.27), then the equilibrium distribution is given by expression (3.28). Indeed the discussion proves this assertion if there exist numbers $N^*(1), N^*(2), \ldots$ such that $\Psi(N(1), N(2), \ldots) = 0$ if $N(i) > N^*(i)$. This restriction can be removed by proving the assertion directly.

Theorem 3.14. *If in a stationary network of quasi-reversible queues the* *arrival rates at the system take the form (3.27) then the equilibrium distribution is given by expression (3.28).*

Proof. The reversed process might consist of the reversed queues with customers of type i entering the system at rate (3.27) and moving backwards through their route. We can use Theorem 1.13 to check that the reversed process does indeed take this form and that (3.28) is the equilibrium distribution. The conditions of Theorem 1.13 are readily verified; they take the same form as before apart from a straightforward embellishment for transitions caused by a customer entering or leaving the system.

To illustrate how we might use Theorem 3.14 let $\mathcal{I}(1), \mathcal{I}(2), \ldots$ be sets of customer types (not necessarily disjoint) and let

$$N(\mathcal{I}) = \sum_{i \in \mathcal{I}} N(i)$$

If Ψ is of the form

$$\Psi(N(1), N(2), \ldots) = \prod_{\mathcal{I}} \prod_{N=0}^{N(\mathcal{I})-1} \psi_{\mathcal{I}}(N)$$

then the arrival rate of customers of type i is

$$\prod_{\mathcal{I}: i \in \mathcal{I}} \psi_{\mathcal{I}}(N(\mathcal{I}))$$

If $\psi_{\mathcal{I}} \equiv 1$ for all $\mathcal{I}$ then customers of type i arrive at the system in a Poisson stream at rate $\nu(i)$, and we have the ordinary open network of queues. By varying the function $\psi_{\{i\}}$ we can make the rate of arrival of customers of type i depend upon the number of such customers already in the system. If $\mathcal{I}$ corresponds to the set of types used to represent a customer whose route is random then it may be appropriate to use the function $\psi_{\mathcal{I}}$ rather than the functions $\psi_{\{i\}}$, $i \in \mathcal{I}$. Of course the function $\psi_{\{1,2,\ldots\}}$ enables the rate of arrival of all customer types to be affected by the total number of customers in the system; for example if we let $\psi_{\{1,2,\ldots\}}(N^*) = 0$ then the system will saturate when the number of customers in it reaches N^*.

In spite of these examples the arrival rates allowed by Theorem 3.14 are of a fairly restricted form. They can depend upon the number of customers of each type in the system but not upon the position within the network of these customers. Further, the dependence upon the number of customers of each type must be expressible in terms of the function Ψ.

We shall end this section with a simple example to illustrate Theorem 3.14.

A repair shop. Consider a repair shop that accepts two types of job. The shop employs K repairmen altogether, of whom K_1 can deal with jobs of type 1, K_2 can deal with jobs of type 2, and the remainder, K_3, can deal with both types of job. The shop is interested in choosing the best arrangement (K_1, K_2, K_3). Suppose that jobs of type 1 and 2 arrive in independent Poisson streams of states ν_1 and ν_2 respectively. When a job arrives it is accepted if there is a repairman free who can deal with it; otherwise the job is lost. Jobs being dealt with can be reshuffled amongst the repairmen if this allows an extra job to be accepted. The time taken to deal with a job of type i is arbitrarily distributed with mean μ_i^{-1} and is not affected by any reshuffling that may be necessary.

Let n_i be the number of jobs of type i in the shop and let

$$\Psi(n_1, n_2) = \begin{cases} 1 & n_1 \le K_1 + K_3,\ n_2 \le K_2 + K_3,\ n_1 + n_2 \le K_1 + K_2 + K_3 \\ 0 & \text{otherwise} \end{cases}$$

The rules we have specified above imply that the probability intensity a job of type i will arrive and be accepted when n_j jobs of type j are already there is

$$\nu_1 \frac{\Psi(n_1+1, n_2)}{\Psi(n_1, n_2)} \qquad i = 1$$

and

$$\nu_2 \frac{\Psi(n_1, n_2+1)}{\Psi(n_1, n_2)} \qquad i = 2$$

The entire system thus behaves as an infinite-server queue with two types of customer whose arrival pattern is of the form (3.27). Theorem 3.14 thus shows that in equilibrium

$$\pi(n_1, n_2) = B\Psi(n_1, n_2) \frac{1}{n_1!} \left(\frac{\nu_1}{\mu_1}\right)^{n_1} \frac{1}{n_2!} \left(\frac{\nu_2}{\mu_2}\right)^{n_2}$$

Exercise 3.5.5 shows that the system appropriately augmented is quasi-reversible and that, counting lost jobs, the departure streams formed by jobs of types 1 and 2 are independent and Poisson.

Exercises 3.5

1. Write down Kolmogorov's criteria for the queue considered in Lemma 3.13. Observe that these conditions on the functions $\phi_c(n(1), n(2), \ldots)$ are equivalent to the existence of a function Φ satisfying equation (3.26).
2. Choose a function Ψ so that the queues considered in Exercise 1.6.1 have arrival rates of the form (3.27).
3. Show that if Ψ takes the value zero unless $N(\mathcal{J}) = N^*(\mathcal{J})$ for each $\mathcal{J}$ then the resulting network is equivalent to the closed network considered in Section 3.4. Observe that an appropriate choice of Ψ produces

a mixed system in which certain types of customer can enter and leave the system while customers of other types can do neither.

4. Show that when a customer of type i arrives at queue $j = r(i, s)$ at stage s of his route the probability that the disposition of the other customers in the system is $(\mathbf{x}_1, \mathbf{x}_2, \ldots, \mathbf{x}_J)$ is proportional to

$$\Psi(N(1), N(2), \ldots, N(i)+1, \ldots)\pi(\mathbf{x}_1)\pi(\mathbf{x}_2)\cdots\pi(\mathbf{x}_J)$$

where $(N(1), N(2), \ldots, N(i), \ldots)$ is calculated from the disposition $(\mathbf{x}_1, \mathbf{x}_2, \ldots, \mathbf{x}_J)$.

5. Suppose that expression (3.27) takes values not greater than $\nu(i)$. Then the apparent arrival rate (3.27) will arise if customers of type i arrive at the system in a Poisson stream of rate $\nu(i)$ but a customer of type i is lost with probability

$$1 - \frac{\Psi(N(1), N(2), \ldots, N(i)+1, \ldots)}{\Psi(N(1), N(2), \ldots, N(i), \ldots)}$$

Show that if I is finite and the state of the system is augmented so that it signals when customers are lost, then the resulting process is quasi-reversible with the class of a customer given by his type.

6. Generalize the model of a repair shop to allow I types of job, with each repairman able to deal with a subset of the I types.
7. Amend the queue considered in Lemma 3.13 so that service requirements at it can be arbitrarily distributed. Show that Lemma 3.13 still holds.
8. Let $\mathbf{x}$ be a quasi-reversible process with equilibrium distribution $\pi(\mathbf{x})$. Suppose now that all transition rates $q(\mathbf{x}, \mathbf{x}')$ which do not correspond to the arrival of a customer are multiplied by $\phi(N)$, where N is the number of customers in the system in state $\mathbf{x}$. Deduce that the resulting system is quasi-reversible with equilibrium distribution

$$B\frac{\pi(\mathbf{x})}{\prod_{l=1}^{N} \phi(l)}$$

either from the equilibrium equations and the partial balance equations (3.11) for the process $\mathbf{x}$, or from Theorem 3.14 and a dilation of the time scale.

9. There are quite subtle ways in which a system with the general arrival rates discussed in this section can be rendered quasi-reversible. Consider a system with three types of customer and with arrival rates of the form (3.27) where $\Psi(0, 0, 0) = \Psi(1, 0, 0) = \Psi(0, 0, 1) = \Psi(0, 1, 1) = 1$, and $\Psi(N(1), N(2), N(3)) = 0$ for other arguments. Then the apparent arrival rate (3.27) could arise as described in Exercise 3.5.5, but if $\nu(1) = \nu(2)$ it could also arise in the following way. Customers of classes 1 and 2 arrive in independent Poisson streams of rates $\nu(1)$ and $\nu(3)$

respectively. When a customer of class 2 arrives he is lost unless the system is empty, in which case he enters it as a customer of type 3. If when a customer of class 1 arrives the system is empty he enters the system as a customer of type 1; if the system contains one customer of type 3 he enters the system as a customer of type 2; otherwise he is lost. Show that, counting lost customers, this system is quasi-reversible with respect to the classes 1 and 2, but not with respect to the types 1, 2, and 3.

10. Consider the following loss priority queueing system with one server. Customers of classes 1 and 2 arrive at a single server in independent Poisson streams of rates ν_1 and ν_2. Customers of class 1 have a higher priority and an arriving customer of class 1 interrupts the service of a customer of class 2; the interrupted service is resumed when the customer of class 1 has his service completed. An arriving customer is lost if a customer of the same or higher priority is being served at the time of arrival. Service times of customers of class i are arbitrarily distributed with mean μ_i^{-1}, $i=1,2$. Show that this system arises when the system considered in the previous exercise contains just a stack. Deduce that the equilibrium probability the system is empty is

$$\frac{\mu_1\mu_2}{(\nu_1+\mu_1)(\nu_2+\mu_2)}$$

Generalize the model to deal with I priority classes.

11. Suppose that the system described in the preceding exercise is amended in the following way: the service time of a customer of type 1 depends upon whether or not he is the only customer in the system. If he is it is arbitrarily distributed with mean μ_1^{-1}; if he has interrupted the service of a customer of type 2 then his service time is arbitrarily distributed with mean μ^{-1}. Show that the equilibrium probability the system is empty is now

$$\frac{\mu_1\mu_2\mu}{\mu(\mu_1\mu_2+\nu_1\mu_2+\nu_2\mu_1)+\nu_1\nu_2\mu_1}$$

CHAPTER 4
Examples of Queueing Networks

In this chapter various examples of queueing networks will be described to illustrate the uses of earlier results and to indicate their limitations.

4.1 COMMUNICATION NETWORKS

An example of a communication network is a telegraph system such as that illustrated in Fig. 4.1(a), where a graph represents the system with vertices and directed edges corresponding to cities and directed channels respectively. Messages are generated in a city and require to be transmitted, possibly via intermediate relay cities, to their destination city. The transmission of message from a city cannot begin until the entire message has been received at that city. Each channel has a maximum capacity and hence the various messages interact. We can model this situation as a network of queues by regarding each message as a customer and each channel as a queue (see Fig. 4.1b). We shall suppose that messages arrive from outside the system (each with its route through the channels of the system) in independent Poisson streams. It is not obvious how the progress of a message through a channel can be represented by a customer passing through a queue, and we shall discuss two possible models.

The time taken for a message to pass along a channel depends on various factors, including the length of the message and random effects associated with the channel. In our first model we shall suppose that the time a message takes to pass along channel j is exponentially distributed with mean ϕ_j^{-1} and independent of the time it takes to pass along other channels along its route. We shall call ϕ_j the capacity of channel j. This model might be appropriate if the random effects associated with a channel are predominant; for example it will arise if the channels are noisy and a message has to be repeatedly transmitted over a channel until it is received without error. With this first model each channel behaves as a single-server queue with exponential service times. The most likely queue discipline is first come first served, but a possible alternative is service in random order. With either of these disciplines the system will behave as a network of queues of the sort discussed in Section 3.1. Thus if n_j is the number of messages waiting at channel j, including the message being transmitted, then in equilibrium

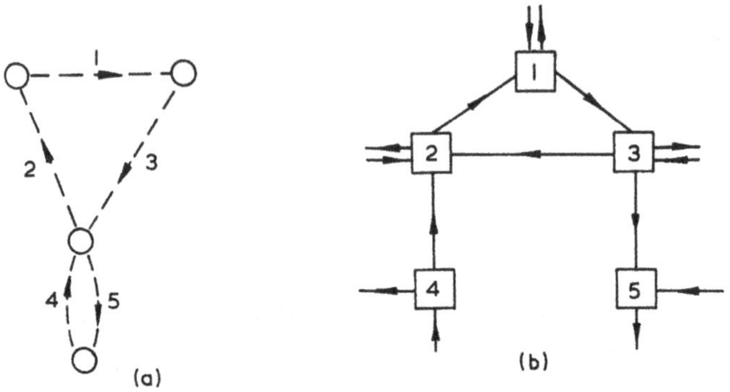

Fig. 4.1 (a) A telegraph system and (b) its representation as a network of queues

$n_1, n_2, \ldots, n_J$ are independent and

$$P(n_j = n) = \left(1 - \frac{a_j}{\phi_j}\right)\left(\frac{a_j}{\phi_j}\right)^n \tag{4.1}$$

where a_j is the average arrival rate at queue j.

In our second model we shall represent a message by a customer whose service requirement is the *same* at every queue along his route, is arbitrarily distributed with unit mean, and is independent from customer to customer. This model might be appropriate if the length of a message is important in determining the time taken to transmit it over a channel. If each channel can be modelled by a symmetric queue then the system will behave as an open network of quasi-reversible queues, with the type of a customer indicating his service requirement as well as his route through the network. The condition that a channel be modelled by a symmetric queue is a severe limitation but allows two queue disciplines which might be appropriate in the context. In the first the channel capacity is divided equally between all the messages waiting for transmission at the channel, corresponding to the server-sharing queue discussed in Section 3.3; this discipline might occur in computer networks and results in short messages being transmitted fairly quickly, even through a congested channel. In the second the channel capacity is devoted entirely to the last arriving message, corresponding to the stack discussed in Section 3.3; this discipline might be a reasonable approximation in a communication network in which the last message sent takes priority over earlier messages. If channel j can supply service effort at rate ϕ_j then in equilibrium $n_1, n_2, \ldots, n_J$ are independent and n_j again has the distribution (4.1).

In practice neither of the models described above might precisely represent the system under scrutiny, but the fact that the same distribution (4.1) emerges under a variety of different assumptions suggests that it might be a good first approximation. We shall now investigate some of its consequences.

The distribution (4.1) implies that the mean number of customers waiting at queue j is $a_j/(\phi_j - a_j)$ and that the average waiting time of a customer at queue j is $1/(\phi_j - a_j)$. Suppose now that we can choose the channel capacities $\phi_1, \phi_2, \ldots, \phi_J$ subject to an overall cost constraint

$$\sum_j f_j \phi_j = F \tag{4.2}$$

How should we allocate the resource F between the competing channels in order to minimize the average time spent in the network by a customer or, equivalently, the mean number of customers in the network? To answer this question we proceed just as in Section 2.4.

Theorem 4.1. *The optimal allocation is*

$$\phi_j = a_j + \frac{\sqrt{a_j f_j}}{\sum_k \sqrt{a_k f_k}} \frac{F - \sum_k a_k f_k}{f_j}$$

Proof. The mean number of customers in the network is

$$\sum_j \frac{a_j}{\phi_j - a_j}$$

and our task is to minimize this subject to the constraint (4.2). Introduce the Lagrange multiplier y and let

$$L = \sum_j \frac{a_j}{\phi_j - a_j} + y\left(\sum_j f_j \phi_j - F\right)$$

Setting $\partial L/\partial \phi_j = 0$ we find that L is minimized by the choice

$$\phi_j = a_j + \sqrt{\frac{a_j}{y f_j}}$$

Substitution of this in constraint (4.2) shows that we should choose

$$\frac{1}{\sqrt{y}} = \frac{F - \sum_k a_k f_k}{\sum_k \sqrt{a_k f_k}}$$

which establishes the result.

There are various other situations where the model discussed in this section and the optimal allocation obtained in Theorem 4.1 might prove helpful. For example the model might be appropriate for a manufacturing job-shop, with customers representing items of work which require to be processed by a number of machines, or a road traffic network, with queues representing bottlenecks in the system.

Exercises 4.1

1. Two cities are connected by two directed channels each of capacity ϕ (Fig. 4.2a). Each channel carries messages which are initiated at rate a. It is proposed that the two channels be replaced by a single channel capable of carrying messages in either direction (Fig. 4.2b). Show that the mean waiting time of a message will be decreased if the capacity of the new channel is greater than $\phi + a$.
2. The two communication networks illustrated in Fig. 4.3 are being considered to link three cities. In the first each of the six channels has capacity ϕ, while in the second each of the three channels has capacity 2ϕ. It is anticipated that the rate at which messages will be sent from one city to another city is a. Show that the mean time a message spends in the system will be less for the first network if $\phi < 3a$.
3. Suppose that the service requirements of a customer at successive queues are not identical but are random variables which may depend upon each other and upon the route of the customer. Show that if the queues are symmetric then distribution (4.1) remains valid, with a_j being the average amount of service requirement arriving at queue j per unit time.
4. Let the routes through the network be labelled and suppose that each unit of time a customer on route i remains in the system costs $g(i)$. Show that the optimal allocation is now

$$\phi_j = a_j + \frac{\sqrt{b_j f_j}}{\sum_k \sqrt{b_k f_k}} \frac{F - \sum_k a_k f_k}{f_j}$$

   where $b_j = \sum_i g(i) a_j(i)$ and $a_j(i)$ is the average arrival rate at queue j of customers on route i.
5. If the queues are symmetric and if the service requirement of a customer on route i may depend upon his type, as in Exercise 4.1.3, show that the optimal allocation remains as in Exercise 4.1.4, but with $a_j(i)$ interpreted as the average rate at which service requirement for customers on route i arrives at queue j.

Fig. 4.2 Alternative communication networks

Fig. 4.3 Alternative communication networks

6. Suppose that service effort is supplied at queue j at rate

$$\phi_j(n)=\frac{n}{n+r-1}\phi_j$$

where r is a positive constant. The case $r=1$ corresponds to the model discussed in this section. Show that if $\phi_1, \phi_2, \ldots, \phi_J$ can be chosen subject to the constraint (4.2) then the optimal allocation is that given in Theorem 4.1.

7. Suppose that service effort is supplied at queue j at rate

$$\phi_j(n)=n\phi_j$$

This might be appropriate if any number of messages can be transmitted at the same time by a channel. Show that if $\phi_1, \phi_2, \ldots, \phi_J$ can be chosen subject to the constraint (4.2) then the optimal allocation is

$$\phi_j=\frac{\sqrt{a_jf_j}}{\sum_k \sqrt{a_kf_k}}\frac{F}{f_j}$$

Show that if $\phi_1, \phi_2, \ldots, \phi_J$ can be chosen subject to the constraint

$$\sum_j \log \phi_j = F$$

then in the optimal allocation $\phi_j/\phi_k = a_j/a_k$.

8. The model of a communication network discussed in this section is, of course, not the only available model, and other models may lead to quite different conclusions. For example suppose arrivals at the series of queues illustrated in Fig. 2.2 form a Poisson process, and the service requirements of customers at queues are all equal to the same fixed value. The model of this section deals with the case where the queues are symmetric, but is inadequate if each queue is a first come first served single-server queue. Show that in this case there will never be more than one customer in queue j, $j=2, 3, \ldots, J$.

4.2 MACHINE INTERFERENCE

The basic form of the machine interference problem is as follows. There are N machines under the care of a single operative. From time to time a machine stops and requires the attention of the operative before it can resume running. The operative can only attend one machine at a time, and so if two or more machines are stopped the others must wait for attention. Thus the machines interfere with one another, and a matter of interest is the extent of this interference.

Suppose initially that the N machines are identical, that the running time of a machine before it stops is exponentially distributed with mean R and that the service time a machine requires from the operative before it can resume running is exponentially distributed with mean S. All running times and service times are assumed independent. The machines that are stopped queue to receive the attention of the operative and the two most common disciplines are for the operative to attend to them in the order of their stopping or in a random order. We can regard the machines that are running as forming a queue also, a queue with an infinite number of servers where a machine remains until it next stops. Observe that service requirements in this queue are in fact machine running times.

The system can thus be represented by the closed queueing network shown in Fig. 4.4. Both queues are quasi-reversible: they would behave in isolation as an $M/M/1$ queue and an $M/M/\infty$ queue respectively. If n_0 is the number of stopped machines and $n_1 = N - n_0$ the number of running machines then in equilibrium

$$\pi(n_0, n_1) = BS^{n_0}\frac{R^{n_1}}{n_1!} \tag{4.3}$$

and from this distribution it is possible to calculate quantities of interest such as the proportion of time the operative or a machine is busy. The above system is in fact a simple example of a closed migration process, and indeed the equilibrium distribution could easily have been determined from the observation that n_1 is a birth and death process. The reason for considering the system as a closed network of quasi-reversible queues is that viewed in this light it becomes readily apparent which of the assumptions underlying the model are crucial in determining the distribution (4.3) and which are unnecessary.

The assumption that running times are exponentially distributed is clearly unnecessary. If running times have a general distribution then the queue containing running machines is equivalent to one which would in isolation

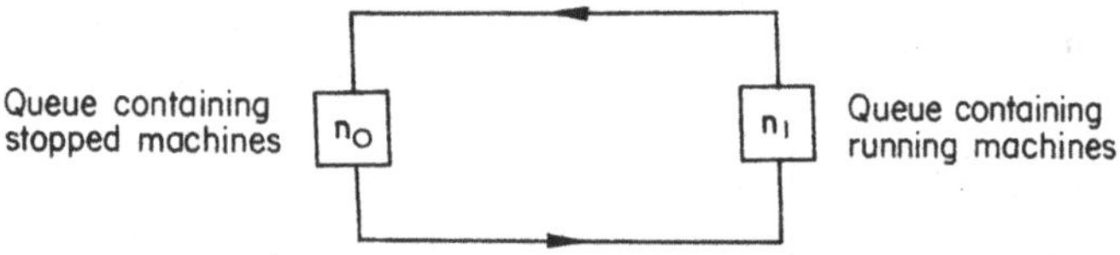

Fig. 4.4 The basic machine interference model

behave as an $M/G/\infty$ queue. Thus the queue remains quasi-reversible and expression (4.3) remains the equilibrium distribution, with R the mean of the running time distribution. The assumption that service times are exponentially distributed cannot be so easily relaxed since the queue containing stopped machines has just one server; with the queue disciplines of interest a single-server will not be quasi-reversible unless service times are exponentially distributed.

In terms of the closed queueing networks of Section 3.4 the present model has just one type of customer whose route consists of two stages with $r(1,1)=0$ and $r(1,2)=1$. We shall now consider the effect on the model of allowing longer routes. Suppose, for example, that the route consists of four stages, with $r(1,1)=0$, $r(1,2)=1$, $r(1,3)=0$, and $r(1,4)=1$. Since queue 1 is a symmetric queue we can allow the service requirement there to have a general distribution and to depend upon whether the customer has reached stage 2 or stage 4 of his route. Let R' and R'' be the mean service requirement at stage 2 and stage 4 respectively. Thus the mean running time of a machine alternates between R' and R'', changing after each service. Since queue 0 is not a symmetric queue we must retain the condition that service times be exponentially distributed, and they must not depend upon the stage reached. Expression (4.3) remains valid, with R defined as the overall mean running time $(R'+R'')/2$ (Exercise 3.4.4). Indeed the result will remain true even if the service requirements at stages 2 and 4 are dependent: in terms of Section 3.4 this corresponds to a customer being allocated a random type each time he begins his route. By fully exploiting the types and routes of the last chapter it is possible to allow a machine's running time to depend upon any number of its previous running times; expression (4.3) will remain valid with R defined as the overall mean running time.

Until now we have supposed that the N machines are identical. When the machines differ it is helpful to view the system as the network shown in Fig. 4.5. Machine i $(i=1,2,\ldots,N)$ remains in queue i while it is running and moves to queue 0 when it stops. After it has been serviced it returns to

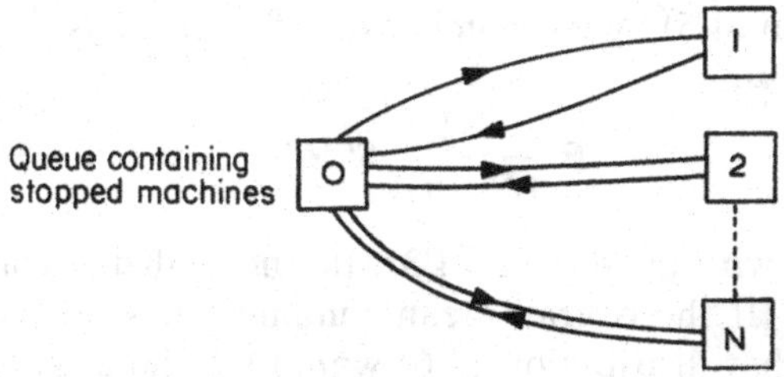

Fig. 4.5 Machine interference with differing machines

queue i. Assume that service times are exponentially distributed with mean S. Initially we shall also assume that the running time of machine i is exponentially distributed with mean R_i, with all running times and service times independent. The state of the system can be written as $(\mathbf{x}_0, x_1, x_2, \ldots, x_N)$ where $\mathbf{x}_0 = (t(1), t(2), \ldots, t(n_0))$ is a listing in order of the machines waiting for service, and $x_j = 1$ or 0 depending on whether machine j is running or not. From Theorem 3.12 we can deduce that the equilibrium distribution is

$$\pi(\mathbf{x}_0, x_1, x_2, \ldots, x_J) = B'S^{n_0} \prod_{i=1}^{N} R_i^{x_i} \tag{4.4}$$

Thus, given that machines $t(1), t(2), \ldots, t(n_0)$ are stopped, each possible ordering of them within queue 0 is equally likely. To obtain the probability that, say, machines $1, 2, \ldots, n$ are stopped and machines $n+1, n+2, \ldots, N$ are running we need to sum expression (4.4) over the $n!$ different orderings of machines $1, 2, \ldots, n$, giving

$$B'n!S^n \prod_{i=n+1}^{N} R_i \tag{4.5}$$

Observe that queue i $(i = 1, 2, \ldots, N)$ is quasi-reversible even when service requirements at this queue are not exponentially distributed: queue i can be considered to be an example of any one of the four symmetric queues described in Section 3.3, since at most one customer is ever present in it. Thus we can generalize the model to allow a machine's running time to have an arbitrary distribution and to depend upon its previous running times. The state of the system will become more complicated since it will need to record more information about each machine. Nevertheless, if the system is in equilibrium the probability that machines $1, 2, \ldots, n$ are stopped and the others running will still be given by expression (4.5), with R_i the overall mean running time of machine i (Exercise 3.4.2). If the overall mean running times of the machines are equal, so that

$$R_1 = R_2 = \cdots = R_N = R$$

then to obtain the probability that n machines are stopped we need to multiply expression (4.5) by the number of different ways n machines can be chosen from N, giving

$$B' \frac{N!}{(N-n)!} S^n R^{N-n} \tag{4.6}$$

This is consistent with expression (4.3); the normalizing constants bear the relation $B = B'N!$ If the overall mean running times are not equal then it might be hoped that distribution (4.6) would still hold with R taken as an average of $R_1, R_2, \ldots, R_N$. In fact this is not true except in the approximate sense explored in Exercise 4.2.3.

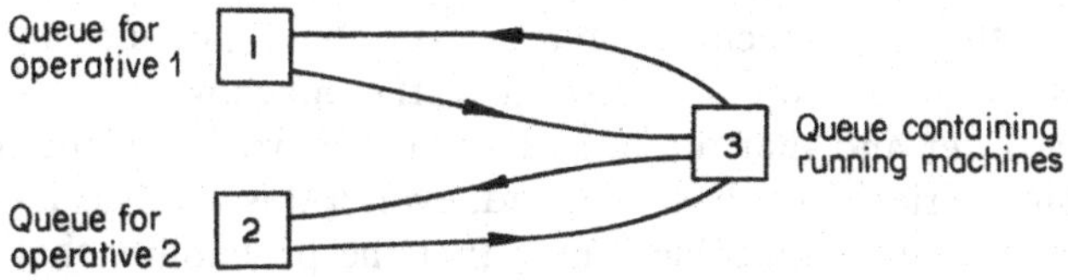

Fig. 4.6 Machine interference with two forms of stoppage

In our final example from the area of machine interference we shall suppose there are N identical machines and that there are two possible reasons for a machine stopping. We shall suppose there are two operatives, one to deal with each form of stoppage. Figure 4.6 illustrates the system; queues 1 and 2 contain machines awaiting service from operatives 1 and 2 respectively and queue 3 contains running machines. Assume that service times at queues 1 and 2 are exponentially distributed with means S_1 and S_2 respectively and that running times are arbitrarily distributed with mean R. Initially we shall also assume that all running times and service times are independent and that when a machine stops it requires attention from operative 1 with probability p_1 and from operative 2 with probability p_2 $(=1-p_1)$ independently of its past history. From Theorem 3.12 it is a simple matter to deduce that in equilibrium

$$\pi(n_1, n_2, n_3) = B(p_1S_1)^{n_1}(p_2S_2)^{n_2}\frac{R^{n_3}}{n_3!} \tag{4.7}$$

As might be expected the above assumptions can be considerably relaxed. If a machine alternately visits operatives 1 and 2 then expression (4.7) remains valid with $p_1 = p_2 = \frac{1}{2}$. If a machine's running time and its reason for stopping at the end of that running time are dependent then we let R be the overall mean running time of a machine. Indeed we can even allow a machine's running time and its reason for stopping to depend upon previous running times and previous reasons for stopping for that machine. Expression (4.7) remains valid with R the overall mean running time of a machine and p_1 the overall proportion of stoppages that require operative 1.

In all of the examples discussed in this section it has been necessary to assume that service times are exponentially distributed and independent of each other and of running times, since the queue for a server is not a symmetric queue. In the next section we shall consider a model closely related to the model of this section, but where it is reasonable to suppose that all the queues involved are symmetric.

Exercises 4.2

1. Deduce from expression (4.3) that n_1 has the same distribution as the number of calls connected in the telephone exchange model of Section 1.3.

2. Consider the basic machine interference model governed by Fig. 4.4 and expression (4.3). Suppose now that the machines are arranged in a priority order and that the operative always works on the machine with the highest priority of those stopped, even if this involves interrupting the service of another machine. Show that the probability that the machine which is kth in the priority order is running is

$$\left[k-(k-1)\frac{f_{k-2}}{f_{k-1}}\right]\frac{1}{f_k}$$

where f_k is given by the recursion

$$f_k = 1+\frac{Sk}{R}f_{k-1}$$

with $f_0 = 1$.

3. Consider the machine interference model with differing machines governed by Fig. 4.5, in which the mean running times of the machines are $R_1, R_2, \ldots, R_N$. Show that if the mean service time S is small a good approximation to the expected number of machines stopped and to the probability that no machines are stopped can be obtained from the basic machine interference model if in that model the mean running time of each machine, R, is given by

$$\frac{1}{R}=\frac{1}{N}\left(\frac{1}{R_1}+\frac{1}{R_2}+\cdots+\frac{1}{R_N}\right)$$

If S is large show that a good approximation to the expected number of machines stopped can be obtained with

$$R=\frac{1}{N}(R_1+R_2+\cdots+R_N)$$

and to the probability that no machines are stopped with

$$R=(R_1R_2\cdots R_N)^{1/N}$$

(A good approximation when S is small or large is one which is accurate to within $o(S)$ or $o(1/S)$ respectively.)

4. Consider the machine interference model illustrated in Fig. 4.5. Suppose now that while the number of stopped machines is n the remaining $N-n$ machines work at a reduced rate $\psi(n)$, and their remaining running times decrease at rate $\psi(n)$ rather than unity. Use Exercise 3.5.8 to find the equilibrium distribution for the system.

5. Consider the machine model with differing machines illustrated in Fig. 4.5. Suppose now that whenever the number of stopped machines reaches M the remaining $N-M$ machines pause, i.e. they cease running

until the operative has finished serving one of the stopped machines and they then resume running where they left off. Deduce from the previous exercise that the equilibrium distribution for the system still takes the form (4.4), but over a smaller state space. Observe that in the case $M=1$ queue 0 is a symmetric queue and so the equilibrium distribution will be the same if the service time of a machine is arbitrarily distributed and dependent on that machine's earlier service and running times provided its overall mean is still S. This system could be viewed as a model of a complex device comprising N units which stops functioning if any one of the units fails.

6. Outline how the machine interference models described can be generalized to allow more than one or two operatives.

4.3 TIMESHARING COMPUTERS

Figure 4.7 illustrates how a queueing network may arise as a much simplified model of a timesharing computer. Queue 0 represents the central processing unit of the computer and queues $1, 2, \ldots, N$ represent terminals. The N customers in the queueing network correspond to jobs, and job i $(i=1, 2, \ldots, N)$ is either being dealt with by the central processing unit or is with the computer user at terminal i. The model as described so far is equivalent to the machine interference model illustrated in Fig. 4.5. The models diverge when we consider the appropriate queue discipline for queue 0. For the present application the most natural assumption is that queue 0 is a single-server queue operating with the server-sharing discipline described in Section 3.3. Queue 0 will then be a symmetric queue and we can allow service requirements at this queue as well as at queues $1, 2, \ldots, N$ to be arbitrarily distributed. Let S_i and R_i be the mean service requirement of customer i at queue 0 and at queue i respectively. Thus S_i and R_i could be

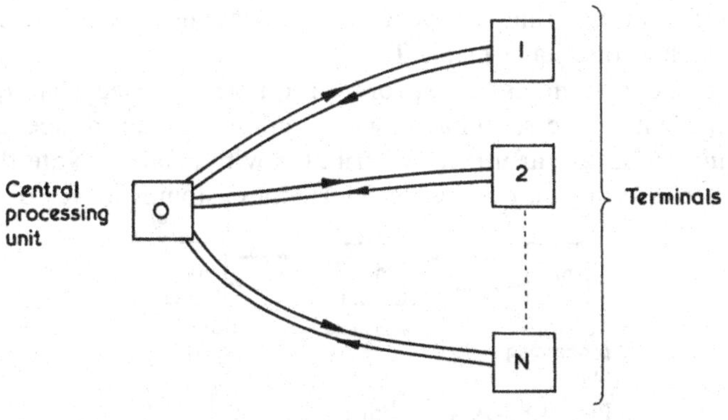

Fig. 4.7 A timesharing computer

called the mean processor requirement and the mean think time respectively for job i. The equilibrium probability that, say, jobs $1, 2, \ldots, n$ are with the central processing unit and jobs $n+1$, $n+2, \ldots, N$ are at terminals is

$$Bn!S_1S_2 \cdots S_nR_{n+1}R_{n+2} \cdots R_N$$

and is insensitive to the form of the distributions involved. This is true even if the service requirements of customer i at queue 0 or i depend upon his earlier service requirements at either or both queues. Note, however, that service requirements of different customers must be independent. The above equilibrium probability allows us to calculate quantities of interest such as the probability that the central processing unit is idle or the proportion of his time a user spends waiting for his job to return to the terminal.

We shall now consider an extension of the model which allows a user, with his job, to leave the terminal. Suppose there is a finite source population of potential users who may wish to use the computer. For simplicity assume these users are identical. Let n_2 be the number of users not using the computer, let n_1 be the number of users at terminals whose jobs are awaiting a response from them, and let n_0 be the number of jobs being dealt with by the central processing unit (Fig. 4.8). Thus if there are in total N users and M terminals then

$$n_0+n_1+n_2=N$$

and (4.8)

$$n_0+n_1 \leq M$$

If one of the users not using the computer attempts to find a terminal and they are all occupied, that is $n_0+n_1=M$, then he tries again later. If he finds a free terminal then he occupies it for a period while his job oscillates between the terminal and the central processing unit. During this period let the total think time have mean R and let the total processor requirement have mean S. Let the time between a user leaving a terminal and next attempting to find one have mean T.

Regarding the system as a network of queues we see that queue 2, containing potential users, behaves as an infinite-server queue at which service requirements are arbitrarily distributed with mean T. Note that there is a capacity constraint on queues 0 and 1: if a customer leaves queue 2 to

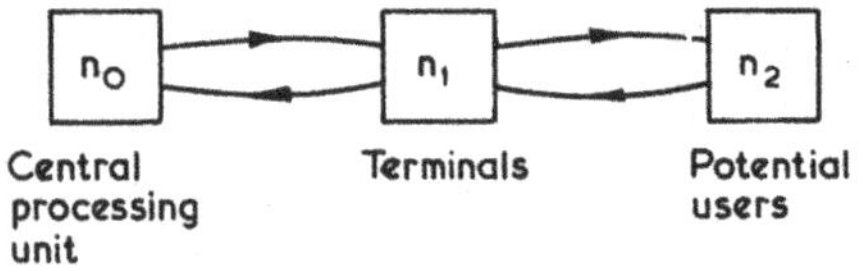

Fig. 4.8 A timesharing computer and its users

find M customers already in queues 0 and 1 he immediately returns to queue 2. Apart from this capacity constraint queue 1 behaves as an infinite-server queue and queue 0 as a server-sharing queue. Hence even with the capacity constraint queues 0 and 1 considered together behave as a quasi-reversible system (Exercise 3.5.5). Thus we can deduce that the equilibrium distribution is

$$\pi(n_0, n_1, n_2) = B_{M,N} S^{n_0} \frac{R^{n_1}}{n_1!} \frac{T^{n_2}}{n_2!} \tag{4.9}$$

over triples (n_0, n_1, n_2) satisfying conditions (4.8). Observe that during the period a user is at a terminal the precise pattern of the oscillations of his job between the terminal and the central processing unit do not affect the result; the distribution (4.9) depends only on the mean quantities R and S. Although it is possible to allow dependencies between the service requirements of a customer at different queues it is not possible to allow the time a user remains away from the terminals to depend upon whether or not he was successful the last time he attempted to find a terminal. This is because a customer entering queue 2 carries with him an indication of his past service requirements but no indication beyond this of his past experience; essentially a customer leaving queue 2 to find M customers already in queues 0 or 1 has his service requirements at these queues met, but instantaneously.

The probability that a user attempting to find a terminal is successful can be determined using part (iii) of Theorem 3.12. It is just the equilibrium probability that queues 0 and 1 contain M customers when there are only $N-1$ customers in the system altogether, and is hence

$$B_{M,N-1} \frac{T^{N-M-1}}{(N-M-1)!} \sum_{n=0}^{M} S^{M-n} \frac{R^n}{n!}$$

Exercises 4.3

1. Extend the model just described to the case where users are not identical.
2. If in the model just described $N, T \to \infty$ with N/T held fixed at ν, check that the equilibrium distribution for (n_0, n_1) becomes

$$\pi(n_0, n_1) = B(\nu S)^{n_0} \frac{(\nu R)^{n_1}}{n_1!}$$

over pairs (n_0, n_1) satisfying $n_0 + n_1 \le M$. Show directly that this is the equilibrium distribution for an open system in which the points in time at which users attempt to find a terminal form a Poisson process.

3. Deduce from Exercise 3.4.6 that for the model illustrated in Fig. 4.7 the mean time a given user spends waiting before his job returns from the central processing unit is proportional to its processor requirement.

Observe that for the model illustrated in Fig. 4.8 the restriction to a given user can be dropped; since the users are identical the mean time taken by any job to pass through the central processing unit is proportional to its processor requirement.

4. In the model illustrated in Fig. 4.8 it was assumed that a customer leaving queue 2 to find M customers in queues 0 and 1 immediately returns to queue 2. Suppose the model is amended in the following way: when the number of customers present in queue 2 drops to $N-M$ the service effort provided at queue 2 becomes zero. Show that the equilibrium distribution is unaltered. Observe that if the time a user spends away from a terminal is exponentially distributed and independent of his experience elsewhere then the two models are equivalent. Unless this assumption is a reasonable one to make, it is unlikely that either model adequately represents the real response of users when all terminals are in use.

4.4 TELETRAFFIC MODELS

We have already discussed models of a telephone exchange in Sections 1.3, 2.1, and 3.3. We shall begin this section by showing how these models can be viewed as special cases of the machine interference models described in Section 4.2. We shall then discuss some further more complicated teletraffic models.

In the simple telephone exchange model of Section 1.3 calls are initiated as a Poisson process, the exchange has K lines, a call initiated when all the lines are busy is lost, and a connected call lasts for an exponentially distributed length of time. This corresponds to the basic machine interference model illustrated in Fig. 4.4 with n_1 the number of busy lines and n_0 the number of idle lines, where $n_0+n_1=K$. Note how the assumptions that calls are initiated as a Poisson process and that calls are lost when all lines are busy correspond to the assumption of exponential service times in the machine interference model. The more general telephone exchange model of Section 3.3 in which call lengths are arbitrarily distributed corresponds to the machine interference model in which running times are arbitrarily distributed. This relationship between telephone exchange models and machine interference models points the way to various generalizations, one of which we will discuss now (others will be considered in Exercises 4.4). Suppose that while a line is busy there is a possibility that it may develop a fault. If this happens the call in progress is allowed to finish, but immediately afterwards the line undergoes repair. We can represent the system by Fig. 4.9 where n_0 is the number of idle lines, n_1 the number of busy lines, and n_2 the number of lines undergoing repair. Suppose that calls are initiated as a Poisson process of rate ν, are independent of each other, and are arbitrarily

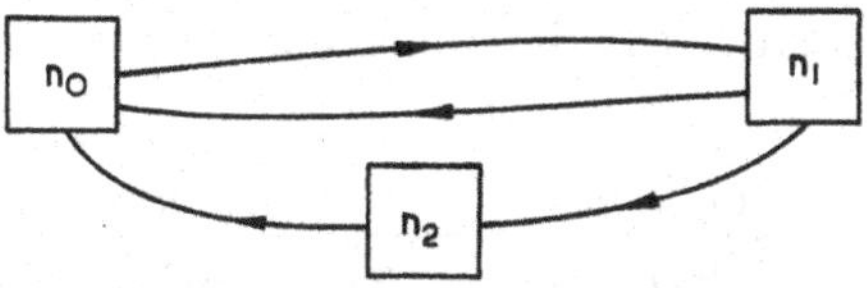

Fig. 4.9 A model of a telephone exchange with unreliable lines

distributed with mean μ^{-1}. Allow the probability intensity that a line develops a fault and the subsequent repair time to depend upon the total time for which the line has been busy and the number of calls it has dealt with since its last repair. Let the mean repair time be λ^{-1} and let m be the mean number of calls a line can handle between successive repairs. Regarding the system as a network of queues we see that queue 0 behaves as a single-server queue at which service requirements are exponentially distributed and queues 1 and 2 behave as infinite-server queues. Observe how the dependence allowed between call lengths, fault occurrence, and repair times is modelled by a dependence between the route and the service requirements at the two symmetric queues of a given customer. In equilibrium

$$\pi(n_1, n_2) = B_K \frac{1}{n_1!}\left(\frac{\nu}{\mu}\right)^{n_1} \frac{1}{n_2!}\left(\frac{\nu}{\lambda m}\right)^{n_2} \qquad 0 \le n_1 + n_2 \le K$$

We shall now consider a teletraffic model which makes simple use of the more general arrival rates discussed in Section 3.5. Suppose there are J exchanges $A_1, A_2, \ldots, A_J$ connected to exchange C via a transit exchange B (Fig. 4.10). Let there be R_j lines between A_j and B and K lines between B and C where $R_j \le K$ and $K < R_1 + R_2 + \cdots + R_J$. Suppose that calls requiring a line between A_j and C are initiated as a Poisson process of rate ν_j and that such calls are lost when all the lines from A_j to B or all the lines from B to C are busy. If a call between A_j and C is connected suppose that the call lasts for an arbitrarily distributed length of time, with mean μ_j^{-1}. Figure 4.10 includes a representation of the system as a network of queues. If n_j is the number of calls in progress between A_j and C for $j = 1, 2, \ldots, J$ then the

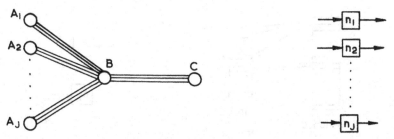

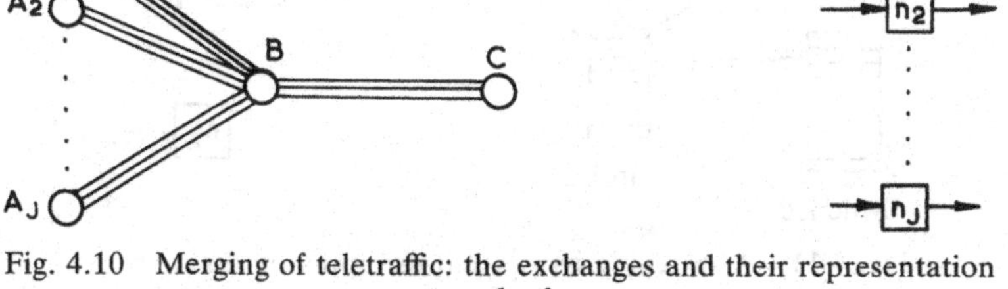

Fig. 4.10 Merging of teletraffic: the exchanges and their representation as a network of queues

probability intensity that a call is connected between A_j and C can be written as

$$\nu_j \frac{\Psi(n_1, n_2, \ldots, n_j+1, \ldots, n_J)}{\Psi(n_1, n_2, \ldots, n_j, \ldots, n_J)}$$

where

$$\Psi(n_1, n_2, \ldots, n_J) = \begin{cases} 1 & \text{if } n_j \leq R_j, j=1, 2, \ldots, J; n_1+n_2+\cdots+n_J \leq K \\ 0 & \text{otherwise} \end{cases}$$

Thus Theorem 3.14 shows that in equilibrium

$$\pi(n_1, n_2, \ldots, n_J) = B \prod_{j=1}^{J} \left(\frac{\nu_j}{\mu_j}\right)^{n_j} \frac{1}{n_j!}$$

$$n_j \leq R_j, j=1, 2, \ldots, J; \ n_1+n_2+\cdots+n_J \leq K$$

Note that in this application of Theorem 3.14 customers of class j visit only queue j.

A more demanding use of Theorem 3.14 will be needed for the following system. A call distributor consists of R_1 switches connected to a first group of K_1 lines and R_2 switches connected to a second group of K_2 lines (see Fig. 4.11), where $R_j \geq K_j$, for $j=1, 2$. Calls are initiated as a Poisson process of rate ν. When a call is initiated it is allocated an idle switch at random, so that if n_j calls are in progress on group j then the probability that the call is routed to group j is $(R_j - n_j)/(R_1+R_2-n_1-n_2)$. A call routed to group j that finds all K_j lines busy is lost. Connected calls last for a time which is arbitrarily distributed with mean μ^{-1}. The probability intensity a call is connected on group j is $\nu(R_j-n_j)/(R_1+R_2-n_1-n_2)$ is $n_j < K_j$ and is zero otherwise. This can be written in the form appropriate for an application of Theorem 3.14, viz. form (3.27), with

$$\Psi(n_1, n_2) = \begin{cases} \dfrac{(R_1+R_2-n_1-n_2)!}{(R_1-n_1)!(R_2-n_2)!} & n_1 \leq K_1, n_2 \leq K_2 \\ 0 & \text{otherwise} \end{cases}$$

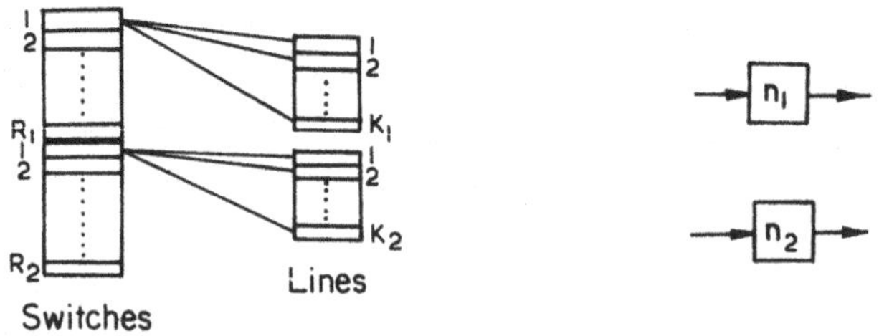

Fig. 4.11 A switching system and its representation as a network of queues

Thus in equilibrium

$$\pi(n_1, n_2) = B\frac{(R_1+R_2-n_1-n_2)!}{(R_1-n_1)!(R_2-n_2)!}\left(\frac{\nu}{\mu}\right)^{n_1+n_2}\frac{1}{n_1!n_2!} \qquad n_1 \le K_1,\ n_2 \le K_2$$

Exercises 4.4

1. Consider a telephone exchange with K lines. Suppose the mean call length on line j is μ_j^{-1}. Show that if the rate at which calls are initiated, ν, is small then a good approximation to the probability a call is lost is given by Erlang's formula
$$\frac{(1/K!)(\nu/\mu)^K}{\sum_{j=0}^{K}(1/j!)(\nu/\mu)^j} \tag{4.10}$$
with $\mu = (\mu_1\mu_2\cdots\mu_K)^{1/K}$, while if ν is large a good approximation is given by the same expression with $\mu = (1/K)(\mu_1+\mu_2+\cdots+\mu_K)$.
2. A switchboard has K lines and one operator. Calls arrive at the switchboard as a Poisson process of rate ν, but calls arriving while all K lines are in use are lost. A call finding a free line has to wait for the operator to answer. The operator deals with waiting calls one at a time and takes an exponentially distributed amount of time with mean λ^{-1} to connect a call to the correct extension, after which the call lasts for an arbitrarily distributed length of time with mean μ^{-1}. Show that the probability k lines are busy is proportional to
$$\left(\frac{\nu}{\lambda}\right)^k \sum_{n=0}^{k}\left(\frac{\lambda}{\mu}\right)^n \frac{1}{n!} \qquad k = 0, 1, \ldots, K$$
3. Consider the model of a telephone exchange with unreliable lines. Show that the probability a call is lost is the same as in the simple telephone exchange model of Section 1.3 if the mean call length in that model is $\mu^{-1}+(\lambda m)^{-1}$.
4. Consider the model of a telephone exchange with unreliable lines. How will the equilibrium distribution be affected if when a fault occurs the call in progress is lost?
5. Consider the model of a telephone exchange with unreliable lines. Suppose that lines with faults must be repaired one at a time and that the time taken to repair a line is exponentially distributed. Show that the probability k lines are busy takes the same form as in Exercise 4.4.2 and discuss the relationship between the two models.
6. Extend the model of the merging of teletraffic to allow calls to be made between exchanges B and C and between exchanges A_j and B. Extend the model to allow more than one transit exchange. Deal with the case where there may be more than one possible route for a call and make the assumption (familiar from the repair shop model of Section 3.5) that

a call is connected if calls already in progress can be reshuffled to make room for it.

7. Extend the model of a switching system to allow J groups of lines. Show that the probability a call is lost is less than in a system in which an arriving call is allocated to group j with probability $R_j/\sum_{i=1}^{J} R_i$. Show that the probability a call is lost is greater than expression (4.10) with $K=\sum_{j=1}^{J} K_j$, unless $R_j=K_j$, $j=1,2,\ldots,J$.
8. The models of this section have assumed that calls are initiated as a Poisson process. Explain how the telephone exchange model considered in Exercise 4.4.2 can be amended to deal with a finite source population of potential callers, as described in Exercise 1.3.5, by allowing idle lines to form a queue with $\phi(n)=N-K+n$ rather then $\phi(n)=1$, for $n>0$, where N is the total size of the source population.
9. In this section we have viewed lines as customers. If there is a finite source population it may be more useful to regard the callers as customers. For example consider a telephone exchange with K lines serving a population of N distinguishable callers. Let μ_i^{-1} be the mean length of a call from caller i, $i=1,2,\ldots,N$, and let λ_i^{-1} be the mean time between the end of his last call (whether connected or lost) and his next attempt to call. Show that in equilibrium the probability callers $i_1, i_2, \ldots, i_n$ are connected is

$$B\prod_{j=1}^{n}\frac{\lambda_{i_j}}{\mu_{i_j}}.$$

Note that in this model intercall times and call lengths can be arbitrarily distributed, but whether a caller is connected or lost must have no effect on his future behaviour.

10. Although in this section we have viewed lines as customers the resulting systems are sometimes quasi-reversible with calls viewed as customers. Consider, for example, the basic telephone exchange model analogous to the machine interference model illustrated in Fig. 4.4. In this model the K lines are regarded as customers and so we can allow the successive call lengths on a given line to be dependent. Suppose now that we view the whole system as a single queue at which calls are the customers. Show that, counting lost calls, this queue is quasi-reversible provided the class of an arriving call does not affect its progress through the queue. Suppose now that this queue and an infinite-server queue form a closed network containing N callers. Since the infinite-server queue is symmetric we can allow the intercall times of a given caller to be dependent. Let λ_i^{-1} be the mean intercall time for caller i, $i=1,2,\ldots,N$, and let μ_k^{-1} be the mean call length for line k, $k=1,2,\ldots,K$. Show that in equilibrium the probability callers $i_1, i_2, \ldots, i_n$

are connected and lines $k_1, k_2, \ldots, k_n$ are busy is

$$B\prod_{j=1}^{n}\frac{\lambda_{i_j}}{\mu_{k_j}}$$

Contrast the patterns of dependence which can be allowed in this model with those that can be allowed in the model of Exercise 4.4.9.

11. Show that in the models of the preceding two exercises the probability a particular caller finds n lines busy when he attempts to make a call is equal to the equilibrium probability that n lines would be busy were the system to contain just the other $N-1$ callers.

4.5 COMPARTMENTAL MODELS

In this section we shall consider systems which have the property that after customers (or particles or individuals) have entered the system they move independently through it. Viewing the system as a network of queues, each queue would in isolation behave as an $M/G/\infty$ queue. The analysis of such systems in equilibrium is fairly straightforward, and there are various applications. We shall mention compartmental models in biology, birth–illness–death processes, and models of manpower systems. Finally, we shall show that in a special case the transient behaviour of the system can be completely described.

In biology compartmental models are used to represent the movement of particles through the various parts of an animal's body. The system is assumed to consist of J compartments with particles entering the system in a Poisson stream and independently moving around between the various compartments before leaving the system. Let ν be the arrival rate of particles at the system and let α_j be the mean time a particle passing through the system spends in compartment j. Then it is a simple consequence of Theorem 3.7 that in equilibrium the number of particles in compartment j is independent of the number of particles in the other compartments and has a Poisson distribution with mean $\nu\alpha_j$. This is true no matter how complicated the motion of the individual particles. For example a particle's stay in a particular compartment may be arbitrarily distributed and may depend upon its past history, as may the compartment the particle chooses to visit next. The essential assumption is that the movements of different particles are independent.

Birth–illness–death processes have the same structure as the above model but the interpretation is rather different. The idea is that an individual is born and passes through various states of health before eventually dying. Given their times of birth the individuals move independently through the J states of the system. The life history of an individual (the states he will pass through and the length of time he will remain in each) is chosen at birth from an arbitrary distribution over all such life histories. If there are N individuals alive let the probability intensity of a birth be $\nu(N)$. The previous model thus corresponds to the case $\nu(N)=\nu$ for all N. Let n_j be the number of individuals in state j and let α_j be the mean time an individual spends in state j throughout its lifetime. Theorem 3.14 shows that when an equilibrium distribution exists it is given by

$$\pi(n_1, n_2, \ldots, n_J) = B\left(\prod_{l=0}^{N-1} \nu(l)\right) \prod_{j=1}^{J} \frac{\alpha_j^{n_j}}{n_j!}$$

Obvious consequences of this are that the equilibrium distribution for N is

$$\pi(N) = B \frac{\alpha^N}{N!} \prod_{l=0}^{N-1} \nu(l)$$

where $\alpha = \sum_{j=1}^{J} \alpha_j$ is the average lifetime of an individual, and that given N the distribution of $(n_1, n_2, \ldots, n_J)$ is multinomial.

The above model could also be used to represent the flow of individuals through a manpower system. The various states would then correspond to grades within the organizational hierarchy. The assumption that individuals move independently of one another prevents the model from dealing with systems in which promotions occur to fill vacancies, rather than when an individual is ready (cf. Exercise 6.3.2). The rate of recruitment $\nu(N)$ will generally be a decreasing function of N, in contrast to the birth–illness–death process where the birth rate $\nu(N)$ will usually be an increasing function of N. As an example of the sort of result which might be useful in this particular application, suppose the distribution function for the total time an individual spends in grade j is $F(u)$. Then Theorems 3.10 and 3.14 show that in equilibrium the amount of experience in that grade which a typical individual there has already acquired has distribution function

$$F^*(x) = \frac{1}{\alpha_j} \int_0^x (1-F(u))\, du$$

Until now we have concerned ourselves with the equilibrium behaviour of compartmental models. If $\nu(N)=\nu$, so that the arriving stream of individuals is Poisson, it is possible to analyse the transient behaviour of the model. Let $p_j(t)$ be the probability that an individual is, a time t after his arrival, in compartment j.

Theorem 4.2. *If the arrival stream is Poisson and if the system is empty at time* 0 *then at time t the number of individuals in compartment j is independent of the number in the other compartments and has a Poisson distribution* *with mean $\nu\alpha_j(t)$ where*

$$\alpha_j(t) = \int_0^t p_j(u)\,du$$

Proof. Let M be the number of arrivals in the interval $(0, t)$. Conditional on M, the instants of arrival $t_1, t_2, \ldots, t_M$ are independent random variables uniformly distributed on $(0, t)$; this follows from the assumption that the arrival process is Poisson. The probability that the arrival at t_r is in compartment j at time t is $p_j(t-t_r)$. Because the individuals move independently we can deduce that

$$E(z_1^{n_1(t)} z_2^{n_2(t)} \cdots z_J^{n_J(t)} \mid M, t_1, t_2, \ldots, t_M) = \prod_{r=1}^{M} \left\{ \sum_j [1 - p_j(t-t_r) + z_j p_j(t-t_r)] \right\}$$

$$= \prod_{r=1}^{M} \left\{ 1 - \sum_j (1-z_j) p_j(t-t_r) \right\}$$

Averaging this over $t_1, t_2, \ldots, t_M$, conditional on M,

$$E(z_1^{n_1(t)} z_2^{n_2(t)} \cdots z_J^{n_J(t)} \mid M) = \prod_{r=1}^{M} \left\{ 1 - \sum_j (1-z_j) t^{-1} \int_0^t p_j(t-u)\,du \right\}$$

$$= \left\{ 1 - \sum_j (1-z_j) t^{-1} \int_0^t p_j(t-u)\,du \right\}^M$$

Averaging over M, which has a Poisson distribution with mean νt,

$$E(z_1^{n_1(t)} z_2^{n_2(t)} \cdots z_J^{n_J(t)}) = \exp\left[-\nu \sum_j (1-z_j) \int_0^t p_j(t-u)\,du \right]$$

$$= \prod_{j=1}^{J} \exp[-(1-z_j)\nu\alpha_j(t)]$$

Hence $n_1(t), n_2(t), \ldots, n_J(t)$ are independent Poisson variables with means $\nu\alpha_j(t)$, which proves the result.

Letting $t \to \infty$ we obtain the previous equilibrium result, since $\int_0^\infty p_j(u)\,du$ is the mean time an individual spends in compartment j throughout its lifetime.

Exercises 4.5

1. In the birth, death, and immigration process considered in Section 1.3 the lifetimes of individuals were exponentially distributed with mean μ^{-1}. Show that the equilibrium distribution (1.14) remains the same if lifetimes are arbitrarily distributed with mean μ^{-1}. Look now at the family size process $(n_1, n_2, \ldots)$ introduced in Section 2.4. By considering

the mean time for which a family has j members alive throughout its entire existence show that the equilibrium distribution for the process $(n_1, n_2, \ldots)$ also remains the same if lifetimes are arbitrarily distributed with mean μ^{-1}. (An alternative proof of this result will be given in Exercise 7.1.9.)

2. In this section we have supposed that individuals are all of the same type. Theorem 3.14 shows that with more than one type of individual more general arrival rates can be allowed. We shall give two examples. Consider two birth–illness–death processes and let $N(i)$ be the number of individuals alive in process i, $i=1, 2$. Find the equilibrium distribution if the birth rate in process i is altered to $(N(i)+1)/(N(1)+N(2)+2)$, $i=1, 2$, and show that although the overall birth rate is constant the total number of individuals alive does not necessarily have a Poisson distribution. Find the equilibrium distribution if the birth rates in processes 1 and 2 are altered to $x^{N(2)}$ and $x^{N(1)}$ respectively $(x<1)$ and show that conditional on the number of individuals alive in one process the number alive in the other process has a Poisson distribution.
3. Consider a birth–illness–death process with two states. Suppose individuals are born into state 1 where they remain for a mean time α_1 and then move to state 2 where they remain for a mean time α_2 before dying. Suppose the birth rate depends only on n_1, the number of individuals in state 1. Use Little's result to find an expression for the mean number of individuals in state 2. Show that it equals $\alpha_2/(1-\nu\alpha_1)$ if the birth rate is $\nu(n_1+1)$.

   Patients arrive at a hospital in a Poisson stream of rate ν, but the hospital redirects them if its K beds are all occupied. An accepted patient stays in the hospital for an average of α_1 days; after he leaves the hospital he attends an outpatients' department for an average of α_2 days. Find the mean number of patients attending the outpatients' department.
4. Cars arrive at the beginning of a long road in a Poisson stream of rate ν from time $t=0$ onwards. A car has a fixed velocity $V>0$ which is a random variable. The velocities of different cars are independent. Show that the number of cars on the first x miles of the road at time t has a Poisson distribution with mean $\nu E[V^{-1}\min\{x, Vt\}]$. What is the distribution of the number of cars between x and y miles along the road at time t?
5. Show that the conclusions contained in Theorem 4.2 are not valid if at time $t=0$ there are already individuals within the system.
6. (Hard) A monkey attempts to climb a tree with a constant positive velocity, but at the points in time of a Poisson process the monkey suffers instantaneous negative displacements, the lengths of which are independent of each other and of the Poisson process, and have a common distribution. The expected net velocity of the monkey is positive. Let $n(s)$ be the number of times the monkey slips backwards past the point s

on the tree, so that $n(s)+1$ is the number of the times the monkey climbs past s in the forward direction. Show that the stochastic process $n(s)$ is identical to the number of individuals alive at time s in a birth, death, and immigration process (Exercise 4.5.1) with the immigration rate equal to the birth rate.

4.6 MISCELLANEOUS APPLICATIONS

Road traffic. Consider an infinitely long road on which there are two kinds of vehicle travelling in the same direction (Fig. 4.12). Some of the vehicles are lorries, travelling at a constant velocity u. The rest of the vehicles are cars which travel at a constant velocity $v(>u)$ unless they are held up behind lorries. Suppose that the cars behind a lorry overtake it one at a time, that the car immediately behind the lorry has to wait for an exponentially distributed period of time before it can overtake, and that these periods are independent with mean μ^{-1}. If we apply a velocity $-u$ to all the vehicles, so that the lorries are reduced to rest, then the lorries can be regarded as single-server queues and the cars as customers with exponential service times at these queues. The gaps between lorries can similarly be regarded as infinite-server queues. If we assume that the points in time at which cars catch up with a given lorry form a Poisson process, then the whole system will behave as an infinite series of quasi-reversible queues. At any given time the positions of the cars not held up behind lorries will form a Poisson process of rate λ_1, say. Let λ_2^{-1} be the mean distance between successive lorries. Each lorry behaves as an $M/M/1$ queue with arrival rate $\lambda_1(v-u)$ and service rate μ, so the mean queue size behind a lorry is

$$\frac{\lambda_1(v-u)}{\mu-\lambda_1(v-u)}$$

provided $\lambda_1(v-u)<\mu$. The average time taken for a car to pass a lorry is $[\mu-\lambda_1(v-u)]^{-1}$ and to catch up with the next lorry is $[\lambda_2(v-u)]^{-1}$. Thus the average speed of a car is

$$u+\frac{\lambda_2^{-1}}{[\mu-\lambda_1(v-u)]^{-1}+[\lambda_2(v-u)]^{-1}}=u+\frac{[\mu-\lambda_1(v-u)](v-u)}{\mu+(\lambda_2-\lambda_1)(v-u)} \quad (4.11)$$

Let ν denote the average rate at which cars pass a given point on the road; then

$$\nu=\lambda_1 v+\frac{\lambda_1\lambda_2(v-u)u}{\mu-\lambda_1(v-u)} \quad (4.12)$$

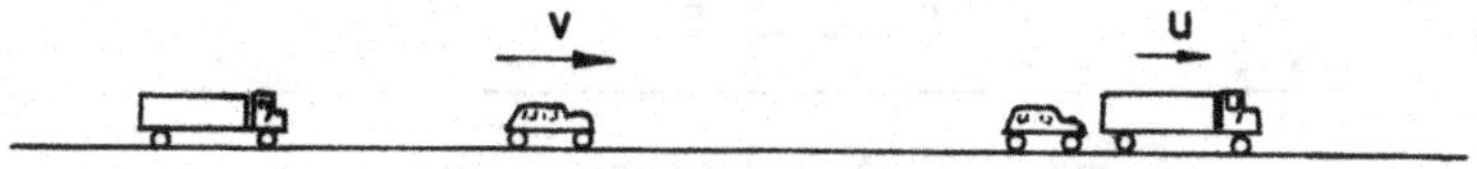

Fig. 4.12 Road traffic

the first term coming from cars passing the point singly and the second from cars passing the point in bunches behind lorries. Expressions (4.11) and (4.12) can be used to explore the effect of varying the parameters of the system. Suppose, for example, that ν varies, with u, v, μ, and λ_2 held constant, corresponding to an increase in the volume of car traffic. Expression (4.12) shows that as λ_1 increases from zero to $\mu/(v-u)$ the parameter ν increases from zero to infinity. Thus for a given value of ν there is a unique solution for λ_1, and using expression (4.11) it can be shown that as ν increases from zero to infinity the average speed of a car decreases from v to u. Other examples are given in Exercises 4.6.

Conveyor belt inspection. Consider a continuously moving conveyor belt carrying items past a quality control office (Fig. 4.13). The office contains K inspectors. When an item reaches the office it enters the office if any of the K inspectors are free—otherwise it continues along the belt. The time taken by an inspector to check an item is arbitrarily distributed with mean μ^{-1}, and after an item has been checked it is replaced on the belt. If items arrive at the office in a Poisson stream at rate ν then this model is equivalent to the queue with no waiting room of Section 3.3, and so the equilibrium probability that j inspectors are occupied is

$$\pi(j) = b_K \left(\frac{\lambda}{\mu}\right)^j \frac{1}{j!} \qquad j = 0, 1, 2, \ldots, K$$

where b_K is the normalizing constant. Further, items pass a point on the conveyer belt downstream from the office in a Poisson stream.

Suppose now that the inspectors are not concentrated in one office but are spread along the conveyor belt as illustrated in Fig. 4.14. Suppose that an item reaching the kth inspector is picked up for checking by that inspector if he is free and if that item has not been checked already by an earlier inspector. It is apparent that in equilibrium the gaps between the inspectors will make no difference to whether or not a particular item is picked up by a given inspector. The stream of items reaching the kth inspector will be the same as if the first $k-1$ inspectors were positioned in a single office. Thus items pass any point along the conveyor belt in a Poisson stream. Despite this, the complete analysis of the system is difficult: for example the probability that the first and third inspectors are both busy will depend on more than just the first moment of the checking time. Exercise 4.6.4 obtains the probability that the kth inspector is busy.

Fig. 4.13 A quality control office

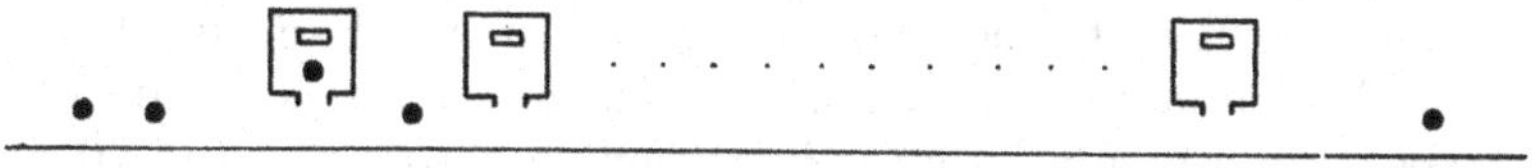

Fig. 4.14 *K* separate offices

If we suppose an item reaching the kth inspector is picked up if he is free and whether or not it has been checked by an earlier inspector, then the stream of items passing any point on the conveyor belt is again Poisson; the system is then a series of quasi-reversible queues.

Electronic counters. A source emits a stream of particles according to a Poisson process of rate ν. An electronic counter is exposed to the stream of particles, but not all the particles are registered. When a particle is registered it causes an aftereffect which lasts for a mean time α. If the counter is suffering from the aftereffects of n particles the probability that an arriving particle will be registered is $p(n)$. Thus n behaves as does the compartmental model of the previous section, and if the counter is in equilibrium the probability that it is suffering from the aftereffects of n particles is

$$\pi(n) = B \frac{(\nu\alpha)^n}{n!} \prod_{l=0}^{n-1} p(l) \tag{4.13}$$

where B is a normalizing constant. The long-run rate at which particles are registered is

$$\nu^* = \nu \sum_{n=0}^{\infty} \pi(n) p(n) \tag{4.14}$$

which can thus be calculated as a function of ν. It is in fact an increasing function of ν (Exercise 4.6.5), and hence from an observed rate ν^* the true rate ν can be determined. For example if $p(0) = 1$ and $p(n) = 0$, $n > 0$ (a type I counter), then expressions (4.13) and (4.14) imply that

$$\nu = \frac{\nu^*}{1 - \nu^* \alpha}$$

Some counters may accept a particle and suffer its aftereffect without registering it. Let $p(n)$ be the probability that a counter accepts an arriving particle when it is suffering from the aftereffects of n particles and let $r(n)$ be the probability that it registers it. Then expression (4.13) again gives the equilibrium probability that the counter is suffering from the aftereffects of n particles, but

$$\nu^* = \nu \sum_{n=0}^{\infty} \pi(n) r(n) \tag{4.15}$$

gives the long-run rate at which particles are registered. Expression (4.15) is not necessarily a monotonic function of ν and so an observed rate ν^* may not correspond to a unique value ν. For example if $r(0)=1$, $r(n)=0$, $n\geq 1$, and $p(n)=1$, $n\geq 0$ (a type II counter), then expressions (4.13) and (4.15) imply that ν is one of the two roots of the equation

$$\nu^* = \nu e^{-\nu\alpha}$$

A garage. A garage employs two mechanics and is fed by a Poisson stream of cars at rate ν. If when a car arrives both mechanics are free the car is equally likely to be assigned to either of the two mechanics. If one mechanic is free the car is assigned to him. If both mechanics are busy the car is lost. The time taken by mechanic i to repair a car has an arbitrary distribution with mean μ_i^{-1}, for $i=1, 2$.

If we view the cars as customers we can obtain the equilibrium distribution for the system from Theorem 3.14. Alternatively, we can view the mechanics as the customers in a closed queueing network. Either approach shows that the equilibrium probability that mechanic i is busy is

$$\frac{\nu(\nu+\mu_i)}{(\nu+\mu_1)(\nu+\mu_2)+\mu_1\mu_2} \tag{4.16}$$

and is insensitive to the form of the repair time distributions. The second approach additionally shows that if the times taken by mechanic i to repair cars form a dependent sequence then this probability is unaltered, with μ_i being the overall mean repair time for mechanic i.

Consider now the stream of cars leaving the garage, including lost cars. Augmenting the state of the network with a flip-flop variable to signal when cars are lost shows that this stream is Poisson and that if we now regard the cars as customers, all of the same class, then the system is quasi-reversible. If there are different classes of cars the system is still quasi-reversible provided the class of a car does not affect its progress through the garage. In this case the class of a car cannot affect its repair time, and hence if the garage is part of a network of quasi-reversible queues a car's repair time cannot depend upon its route or its service requirements at the symmetric queues in the network. In this respect the system is similar to the queues considered in Section 3.1 rather than a symmetric queue, even though its operation involves arbitrary distributions and its equilibrium distribution exhibits a form of insensitivity.

Is it possible to allow the class of a car to affect its progress through the garage? We shall now show that it is. Suppose there are J classes of car. Consider the closed queueing network illustrated in Fig. 4.15; the customers in this network are the two mechanics. The presence of a mechanic in queue 0 indicates that he is idle. The presence of mechanic 1 (respectively 2) in queue jA (respectively jB) indicates that he is repairing a car of class j,

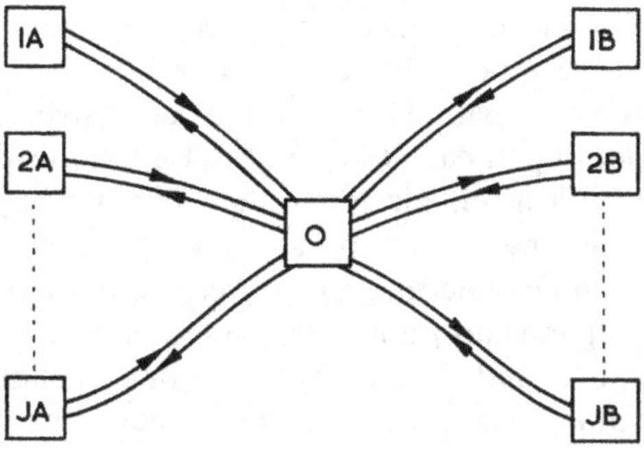

Fig. 4.15 Representation of a garage

$j = 1, 2, \ldots, J$. Cars of class j arrive at the garage in a Poisson stream at rate ν_j, $j = 1, 2, \ldots, J$, and these streams are independent: the implications of this for the queueing network illustrated in Fig. 4.15 are that queue 0 would in isolation behave as an $M/M/1$ queue and that when mechanic 1 (respectively 2) leaves queue 0 he goes to queue jA (respectively jB) with probability ν_j/ν, where $\nu = \sum \nu_j$, independently of the previous history of the system. This determines the routing behaviour of the mechanics. The assumption that if both mechanics are idle an arriving car is equally likely to be assigned to either of them is compatible with the queue disciplines allowed in Section 3.1, as are the alternative assumptions that the car is assigned to the mechanic who has been idle the longest, or the shortest, time. Note that which mechanic is assigned the car cannot depend on the car's class. Queues jA, jB, $j = 1, 2, \ldots, J$, are symmetric queues—they can each contain at most one customer. The service requirement of mechanic 1 at queue jA can depend upon j, upon his previous route, and upon his service requirements at queues 1A, 2A, ..., JA. Thus the repair time of a car of class j assigned to mechanic 1 can depend upon the class of the car, upon the classes of the cars previously repaired by mechanic 1, and upon the repair times of these cars. The equilibrium probability that mechanic i is busy is given by expression (4.16) where μ_i is the overall mean repair time for mechanic i.

There is an important difference between the above process and the process obtained from it by time reversal. In the original process the probability that an arriving car is of class j is independent of the state of the process, but the time taken to repair the car may depend upon the state of the process, in particular upon the class of the car previously repaired by the mechanic assigned to the car. Thus in the reversed process the probability that on leaving queue 0 mechanic 1 goes to queue jA may depend upon the state of the process, in particular upon the time spent by mechanic 1 on his last excursion from queue 0. Hence the system will not in general be quasi-reversible with respect to the J classes of cars.

Let us now restrict the dependencies allowed in the queueing network illustrated in Fig. 4.15. Suppose that each time a car is assigned to mechanic i he is allocated a repair capability, where the successive values allocated to him may be dependent upon each other but not upon the classes of the cars. Let the time spent by a mechanic repairing a car (the repair time) have a distribution determined by the repair capability of the mechanic and the class of the car (we can imagine that a car of class j has a repair requirement whose distribution depends on j and that the repair time is a function of the car's repair requirement and the mechanic's repair capability). Queues jA, jB, $j = 1, 2, \ldots, J$, thus behave as symmetric queues at which the service requirement of a customer has a distribution determined by the corresponding repair capability of the mechanic. The system just described is a restricted form of the system previously discussed, and so the equilibrium probability that mechanic i is busy is still given by expression (4.16). Observe, though, that while the time taken to repair a car may depend upon previous repair times through the sequence of repair capabilities of a mechanic it cannot depend on the classes of the cars previously repaired by him. Hence the reversed process has the property that when mechanic 1 (respectively 2) leaves queue 0 the probability that he goes to queue jA (respectively jB) is ν_j/ν, independently of his past experience and hence of the state of the process. This is enough to show that the system, appropriately augmented to signal lost cars, is quasi-reversible with respect to the J classes of car.

If the garage is part of a network of quasi-reversible queues then the patterns of dependence which can emerge take an interesting form. The time taken by a mechanic to repair a car can be dependent upon the mechanic's experience elsewhere because it can depend on his repair capability, and it can be dependent on the car's experience elsewhere because it can depend on the car's class. However, the experience elsewhere of the car (respectively the mechanic) can depend on that particular repair time only through its class (respectively his repair capability).

Exercises 4.6

1. Consider the road traffic model discussed in this section. Allow ν to vary, with u, v, and μ held fixed and with λ_2 held equal to ν/ku; this corresponds to varying the overall volume of traffic with the ratio of cars to lorries held fixed at k to 1. Show that as ν increases from zero to infinity the average speed of a car decreases from v to u and the mean queue size behind a lorry increases from zero to k.
2. Suppose that in the road traffic model discussed in this section there are only finitely many lorries. In this case the volume of car traffic will be best measured by $\nu = \lambda_1 v$. Investigate the effect of varying the velocity

of the cars on T, the mean time taken by a car to pass the string of lorries, with u, μ, λ_2, and ν held fixed. Show that

(a) If $\mu > \nu$ then as v tends to infinity the mean overtaking time T tends to a finite limit.

(b) If $\mu < \nu$ then as v increases from u to $\nu u/(\nu - \mu)$ the mean overtaking time T decreases from infinity to a minimum and then increases back to infinity.

3. Show how the model of road traffic discussed in this section can be extended to allow cars to travel with different velocities, with cars overtaking each other freely.
4. Show that the probability the kth inspector in the sequence illustrated in Fig. 4.14 is busy is equal to the probability calculated in Exercise 4.2.2, with R/S replaced by λ/μ.
5. Observe that the mean of the distribution (4.13) is an increasing function of ν. Deduce that the rate of registrations ν^* given by expression (4.14) is an increasing function of the rate of arrivals ν.
6. Consider a variant of the type I counter in which $p(0) = p$, $p(n) = 1$, $n > 0$. Show that

$$\nu^* = \frac{\nu p}{p + (1-p)e^{-\nu\alpha}}$$

Consider a variant of the type II counter in which $r(n) = r^n$, $n \geq 0$. Show that

$$\nu^* = \nu e^{-\nu\alpha(1-r)}$$

7. For the garage represented in Fig. 4.15 show that if mechanic i is busy the probability he is repairing a car of class j is

$$\frac{\nu_j \mu_i}{\nu \mu_{ji}}$$

where μ_{ji}^{-1} is the overall mean repair time of a class j car with mechanic i. Show that all the results obtained in this section for a garage with two mechanics can be generalized to a garage with N mechanics.

8. Generalize the model described in Exercise 4.4.10 to allow a call length to depend upon the caller and the line, and discuss the extent to which it can depend upon the previous experience of each of them.
9. A garage employs two mechanics and is fed by two independent Poisson streams of cars. If when a car from the ith stream arrives at the garage the ith mechanic is free he repairs it; if he is busy and the other mechanic is free then the other mechanic repairs it; if both are busy the car is lost. The time taken to repair a car from the ith stream has an arbitrary distribution, for $i = 1, 2$. Use Erlang's formula and Little's result to find the probability the ith mechanic is busy. Observe that it is

insensitive to the form of the repair time distributions and depends only on their means. Observe that the stream of cars leaving the garage, counting lost cars, is Poisson.

10. Suppose the system of the previous exercise is represented by a Markov process $\mathbf{x}(t)$. Show that if from $\mathbf{x}(t_0)$ it is possible to deduce which mechanics are busy at time t_0 then $\mathbf{x}(t_0)$ is *not* independent of the departure process prior to time t_0. Thus if information about which mechanics are busy is included in the state then the system is not quasi-reversible.
11. A number of trucks and excavators are involved in an earthmoving operation. Trucks are loaded by the excavators at the site, after which they travel to a dump, unload the earth, and return to the site. Describe the various ways in which the operation could be modelled as a closed network of quasi-reversible queues.
12. A fleet of vessels operates between a number of loading and discharge ports. Describe the various ways in which the system could be modelled as a closed network of quasi-reversible queues.

CHAPTER 5

Electrical Analogues

In Section 1.4 we discussed the relationship between the approach to equilibrium of a random walk satisfying the detailed balance conditions and the diffusion of charge through an electrical network. We shall begin this chapter by discussing a different aspect of this connection between reversible random walks and electrical networks. Then, in Section 5.2, we shall use the connection to analyse a model representing flow through a network. This flow model is fairly limited in scope but its main interest lies in the fact that it permits blocking, the phenomenon whereby whether an individual leaves a colony is affected by the number of individuals in the colonies to which he may move. The structure of the migration processes and queueing networks discussed in earlier chapters ruled out blocking (although in a very limited way it could be imitated—see Exercise 4.3.4). In general the phenomenon of blocking makes analytical progress difficult; the flow model of this chapter and the reversible migration processes of the next, although highly specialized systems, at least permit blocking and yet remain tractable.

The method we shall use to analyse the flow model can also be applied to an interesting invasion model, and this will be the subject of Section 5.3.

5.1 RANDOM WALKS

Let λ_{jk} be the transition rate from state j to state k of a Markov process with a finite state space G. Let α_j, $j \in G$, be a positive solution of the equations

$$\alpha_j \sum_k \lambda_{jk} = \sum_k \alpha_k \lambda_{kj} \qquad j \in G \tag{5.1}$$

Assume as usual that the transition rates λ_{jk}, $j, k \in G$, define an irreducible Markov process, and hence that the solution to equations (5.1) is unique up to a multiplying factor. We shall regard the Markov process as defining the position of a particle performing a random walk on the graph with vertex set G and with an edge joining $j, k \in G$ if either λ_{jk} or λ_{kj} is positive. If

$$\alpha_j \lambda_{jk} = \alpha_k \lambda_{kj} \qquad j, k \in G \tag{5.2}$$

then we will call the random walk reversible. Thus in this chapter we view reversibility as a property of the transition rates of the Markov process rather than, as in earlier chapters, a property of the equilibrium behaviour of the Markov process. If

$$\lambda_{jk} = \lambda_{kj} \qquad j, k \in G$$

then we will call the random walk symmetric. A symmetric random walk allows $\alpha_j = 1$, $j \in G$, as a solution to equations (5.1) and (5.2), and hence is reversible.

Suppose that observation of the random walk is stopped when the particle first reaches a set $V \subset G$ and that a payment of v_i is received if observation is stopped when the particle arrives at vertex $i \in V$. Let p_j be the expected final payment if the particle starts from vertex j. Considering where the next step of the random walk will take the particle leads to the equation

$$p_j = \sum_k \frac{\lambda_{jk}}{\sum_l \lambda_{jl}} p_k \qquad j \in G - V$$

Thus

$$0 = \sum_k \lambda_{jk}(p_k - p_j) \qquad j \in G - V \tag{5.3}$$

and

$$p_j = v_j \qquad j \in V \tag{5.4}$$

Equations (5.3) and (5.4) have a unique solution for p_j, $j \in G$. Now suppose that the random walk is reversible and define the (possibly infinite) quantity

$$r_{jk} = (\alpha_j \lambda_{jk})^{-1}$$

Thus $r_{jk} = r_{kj}$. In this case equations (5.3) and (5.4) can be rewritten as

$$0 = \sum_k \frac{p_k - p_j}{r_{jk}} \qquad j \in G - V \tag{5.5}$$

and

$$p_j = v_j \qquad j \in V \tag{5.6}$$

But equations (5.5) and (5.6) are precisely Kirchhoff's equations for an electrical network with node set G in which we interpret p_j as the electrical potential of node j, and where nodes j and k are connected by a wire of resistance r_{jk}, and node j is held at potential v_j, $j \in V$. Equation (5.5) expresses the fact that the total current flowing into node j is zero.

The last paragraph dealt with the infinite horizon case in the sense that the particle was allowed as long as necessary to reach the set V. In this paragraph we will suppose there is a finite horizon at time T and that no payment is received if the particle has not reached the set V by time T. Let $p_j(t)$ be the expected final payment if the particle is at vertex j at time $T - t$, with a time t to go before the horizon. Considering the possible events in an interval of time δt leads to the equation

$$p_j(t+\delta t)=\sum_k \lambda_{jk}\,\delta t\, p_k(t)+\left(1-\sum_k \lambda_{jk}\,\delta t\right)p_j(t)+o(\delta t) \qquad j\in G-V$$

Hence

$$\frac{dp_j(t)}{dt}=\sum_k \lambda_{jk}(p_k(t)-p_j(t)) \qquad j\in G-V \tag{5.7}$$

and

$$p_j(t)=v_j \qquad j\in V \tag{5.8}$$

with the initial condition

$$p_j(0)=0 \qquad j\in G-V \tag{5.9}$$

Equations (5.7) are called the backward equations, in contrast to the forward equations (1.16). Equations (5.7), (5.8), and (5.9) have a unique solution for $p_j(t)$, $j\in G$, and

$$\lim_{t\to\infty} p_j(t)=p_j$$

where p_j, $j\in G$, is the solution to equations (5.3) and (5.4). If the random walk is reversible, equations (5.7), (5.8), and (5.9) become

$$\alpha_j\frac{dp_j(t)}{dt}=\sum_k \frac{p_k(t)-p_j(t)}{r_{jk}} \qquad j\in G-V \tag{5.10}$$

and

$$p_j(t)=v_j \qquad j\in V \tag{5.11}$$

with the initial condition

$$p_j(0)=0 \qquad j\in G-V \tag{5.12}$$

Equations (5.10), (5.11), and (5.12) have an electrical interpretation: if we amend the electrical network described in the last paragraph by connecting each node $j\in G-V$ to earth through a capacitor with capacitance α_j and if the potential of each node $j\in G-V$ is zero at time $t=0$, then $p_j(t)$ will be the potential of node j at time t. Observe that as far as the resistors and capacitors are concerned the electrical network is the same as the one described in Section 1.4. Note, however, that in Section 1.4 the analogy was based on the forward equations (1.16) and the correspondence was between charge and probability, both represented by the variable $u_j(t)$. Here the analogy is based on the backward equations (5.7) and the correspondence is between potential and expected final payment, both represented by the variable $p_j(t)$. Of course the expected final payment $p_j(t)$ can itself be used to represent a probability. For example if $V=\{0, 1\}$, $v_0=0$, $v_1=1$, then $p_j(t)$ will be the probability that a random walk starting at vertex j reaches vertex 1 before it reaches vertex 0 and does so before a time t has elapsed.

The analogy of this section can be extended further: we will give two examples.

Example 1. Suppose the payment received is $v_i(t)$ if the particle reaches a vertex $i \in V$ with a time t to go before the horizon. Equation (5.11) will become

$$p_j(t)=v_j(t) \qquad j \in V$$

and in the electrical analogue we will require that node $j \in V$ be maintained at the time-varying potential $v_j(t)$.

Example 2. Suppose a payment of w_j is received if the particle has not reached the set V by the horizon and is at vertex $j \in G-V$ at time T. Then the initial condition (5.12) will become

$$p_j(0)=w_j \qquad j \in G-V$$

and in the electrical analogue we will require that the potential of node $j \in G-V$ at time $t=0$ be w_j.

Exercises 5.1

1. Show that equations (5.10), (5.11), and (5.12) have an alternative electrical interpretation, related to the one given, but with resistors replaced by inductors and capacitors replaced by resistors.
2. Observe that the transition rates λ_{jk}, $j \in V$, $k \in G$, do not affect the expected final payment $p_j(t)$. Deduce that the electrical analogy of this section can still be developed if conditions (5.1) and (5.2) are replaced by the weaker condition

$$\alpha_j\lambda_{jk}=\alpha_k\lambda_{kj} \qquad j, k \in G-V$$

5.2 FLOW MODELS

Consider the following flow model. There are $J-1$ sites (or colonies) labelled 2, 3, . . . , J, and no site may contain more than one individual. If site j is occupied and site k is empty then with probability intensity λ_{jk} the individual at site j moves to site k. If site j is occupied then with probability intensity μ_j the individual at site j leaves the system entirely. If site k is empty then with probability intensity ν_k an individual arrives at site k from outside the system.

In this section we shall analyse this flow model under the assumption that the λ_{jk} are symmetric:

$$\lambda_{jk} = \lambda_{kj} \tag{5.13}$$

(In Section 6.3 we shall consider the model with a different restriction.) We shall suppose that at time $t = 0$ the system is empty and we shall be concerned to find $p_j(t)$, the probability that at time t site j is occupied.

Now consider the following button model. There are $J+1$ sites, labelled $0, 1, \ldots, J$ and each site contains a button. The buttons are distinguishable—we can imagine them to be of different colours. The buttons occupying sites j and k interchange positions with probability intensity λ_{jk} for $j, k = 0, 1, \ldots, J$, where

$$\left.\begin{aligned} \lambda_{0j} &= \lambda_{j0} = \mu_j \\ \lambda_{1j} &= \lambda_{j1} = \nu_j \end{aligned}\right\} \quad j = 2, 3, \ldots, J$$

and

$$\lambda_{01} = \lambda_{10} = 0$$

We see that any particular button performs a symmetric (and hence reversible) random walk around the sites of the system. Now imagine that from time $t = 0$ onwards a button leaving site 1 is painted black and a button entering site 0 is painted white. If $A(t)$ is the set of sites which contain a black button at time t then $A(t)$ behaves stochastically just as does the set of occupied sites in the flow model. Thus to find $p_j(t)$ we need only look backwards through time at the movements of the button which occupies site j at time t. These movements form a symmetric random walk starting from site j with transition intensities λ_{jk} for $j, k = 0, 1, \ldots, J$; $p_j(t)$ is equal to the probability that this random walk reaches site 1 within a time t and does so without passing through site 0. Thus $p_j(t)$ is equal to the potential at time t of node j in an electrical network constructed as follows: join nodes j and k, where $j, k = 0, 1, \ldots, J$, by a wire of resistance λ_{jk}^{-1} whenever $\lambda_{jk} > 0$, and connect nodes $2, 3, \ldots, J$ to earth through a unit capacitor; let the potentials of every node at time $t = 0$ be zero and from time $t = 0$ onwards hold nodes 0 and 1 at potentials 0 and 1 respectively. If at time $t = 0$ site j in the flow model is occupied then the electrical analogy will still hold but we will require that at time $t = 0$ the potential of node j be 1. As $t \to \infty$, $p_j(t) \to p_j$ where p_j is the equilibrium potential of node j in the network; of course, in equilibrium the potentials will be unchanged if all the capacitors are removed.

The average net flow of individuals from site j to site k at time t is

$$\begin{aligned} &\lambda_{jk} \operatorname{Prob}\{\text{site } j \text{ occupied and site } k \text{ empty at time } t\} \\ &\qquad - \lambda_{kj} \operatorname{Prob}\{\text{site } j \text{ empty and site } k \text{ occupied at time } t\} \\ &\qquad\qquad = \lambda_{jk}(p_j(t) - p_k(t)) \end{aligned}$$

which is precisely the current flowing from node j to node k at time t in the electrical network. The average flow of individuals into, and out of, the system at time t are respectively

$$\sum_j \nu_j(1-p_j(t))$$

and

$$\sum_j \mu_j p_j(t)$$

which correspond respectively to the current flowing into, and out of, the electrical network at time t.

For the flow model we have shown that $p_j(t)$, $j=2,3,\ldots,J$, satisfy a set of linear differential equations of the form (5.10), (5.11), and (5.12). Now the set of occupied sites in the flow model is a Markov process and hence forward equations could be deduced which would also form a set of linear differential equations. What have we achieved? Well, first there are only $J-1$ equations in the set we have obtained while there would be 2^{J-1} forward equations. Second, the electrical analogy gives considerable insight into the behaviour of flow models. Of course, a solution to the forward equations would give much more than just a solution for $p_j(t)$, $j=2,3,\ldots,J$; for example it would give the probability that at time t sites j and k are *both* occupied. In fact a solution to the forward equations can be built up inductively starting from a solution for $p_j(t)$, $j=2,3,\ldots,J$ (see Exercise 5.2.1). If we are interested in the equilibrium behaviour of the flow model then the solution to equations (5.5) and (5.6) will give p_j, the equilibrium probability that site j is occupied. An important feature of the flow model which puts it in sharp contrast with the open network models considered in earlier chapters is that the states of different sites are not in general independent. The solution p_j, $j=2,3,\ldots,J$, does not therefore completely determine the equilibrium distribution for the system. It does nevertheless give the most important features of the equilibrium distribution and once again joint probabilities can be built up inductively (Exercise 5.2.2).

The flow model can be extended to allow a site to contain more than one individual. Specifically, suppose that site $j(j=2,3,\ldots,J)$ may contain up to N_j individuals and let $n_j(t)$, $0\le n_j\le N_j$, be the number of individuals at site j at time t. Further suppose that an individual moves from site j to site k with probability intensity

$$\lambda_{jk}\frac{n_j}{N_j}\frac{N_k-n_k}{N_k} \tag{5.14}$$

an individual leaves the system from site j with probability intensity

$$\mu_j \frac{n_j}{N_j} \tag{5.15}$$

and an individual arrives at site k from outside the system with probability intensity

$$\nu_k \frac{N_k - n_k}{N_k} \tag{5.16}$$

This extended model can be obtained as a limiting case of the original flow model; we simply replace site j by N_j fictitious subsites (each of which can contain at most one individual) with infinite intensities of movement between the subsites replacing site j. In this way it is easy to see that if

$$p_j(t) = \frac{E(n_j(t))}{N_j}$$

then $p_j(t)$ is the potential at time t in an electrical network constructed as follows: for $j, k = 2, 3, \ldots, J$ join nodes j and k by a wire of resistance λ_{jk}^{-1} whenever $\lambda_{jk} > 0$, join node j to nodes held at potentials 0 and 1 through wires of resistance μ_j^{-1} and ν_j^{-1} respectively, and connect node j to earth through a capacitor with capacitance N_j. The initial conditions for the electrical network will, as usual, be obtained from the initial conditions for the flow model. It is interesting to note that in the button model corresponding to this extended flow model a button performs a reversible, but not necessarily symmetric, random walk.

Unlike the network models considered in earlier chapters the flow model has few applications. The difficulty is that the symmetry condition (5.13) is too restrictive. It implies that the individuals have no innate tendency to move in any given direction and that flow of individuals through the system is the result of them being forced in at some sites and removed from other sites. This is unlikely to be the case in any of the applications discussed in Chapter 4; in a communication network, for example, we would expect a message to have a preferred direction of travel. We can of course regard the flow model as a naive description of the mechanism governing the movement of electrons in a conductor, and it then provides a physical explanation of the mathematical relationships between random walks and electrical networks obtained in Section 1.4 and the previous section.

Exercises 5.2

1. Suppose we have the solution for $p_j(t)$, $j = 2, 3, \ldots, J$, and that we are interested in finding the joint probabilities that given pairs of sites are occupied at time t. Show that these probabilities correspond to potentials

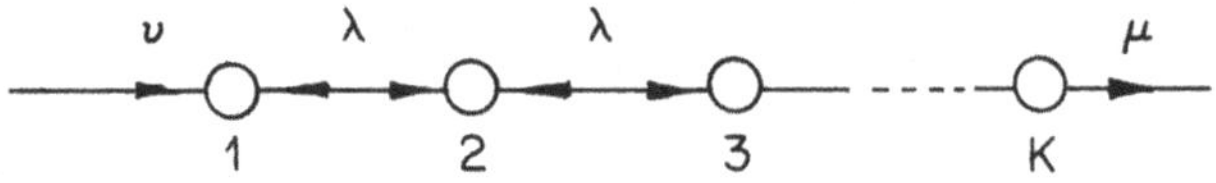

Fig. 5.1 A one-dimensional flow model

in an electrical network with $\frac{1}{2}J(J-1)$ nodes, $J-1$ of which are held at the time-varying potentials $p_j(t)$, $j=2, 3, \ldots, J$.

2. In the one-dimensional flow model illustrated in Fig. 5.1 jumps take place between adjacent sites at rate λ, particles arrive at site 1 at rate ν, and leave from site K at rate μ. Deduce from the electrical analogue that in equilibrium the mean rate of flow of particles through the system is

$$[(K-1)\lambda^{-1}+\nu^{-1}+\mu^{-1}]^{-1}$$

If $\nu=\mu=\lambda$ show that sites j and $k (j<k)$ are both occupied with probability

$$\frac{(K-j)(K+1-k)}{K(K+1)}$$

and both empty with probability

$$\frac{j(k-1)}{K(K+1)}$$

3. Let $A(t)$ be the set of occupied sites in a flow model. In general the reversed process $A(-t)$ is a complicated Markov process quite unlike the original process $A(t)$. Show that in the reversed process obtained from the one-dimensional flow model illustrated in Fig. 5.1 the probability intensity that a particle leaves the system from site 1 depends not only on whether or not a particle is present at site 1 but also upon which of sites $2, 3, \ldots, K$ are occupied.

5.3 INVASION MODELS

Consider the following invasion model. There are $J-1$ sites (labelled $2, 3, \ldots, J$) and each site is coloured either black or white. If sites j and k are different colours then with probability intensity λ_{jk} site k invades site j—when this happens site j takes on the color of site k while site k remains the same colour. If site j is white (respectively black) then with probability intensity ν_j (respectively μ_j) it is invaded from outside the system and becomes black (respectively white). Viewed as a representation of competition between two opposing species or armies the important characteristic of the model is that the chance site j is overrun by site k depends only upon λ_{jk} and not upon the colours involved. At least initially we shall not require that the λ_{jk} be symmetric. We shall suppose that at time $t=0$ all sites in the system are white and we shall be interested in finding $p_j(t)$, the probability that at time t site j is black.

We can replace invasions from outside the system by adding two sites (labelled 0 and 1) which at time $t=0$ are white and black respectively, with

$$\begin{aligned} \lambda_{j0} &= \mu_j & \lambda_{0j} &= 0 \\ \lambda_{j1} &= \nu_j & \lambda_{1j} &= 0 \end{aligned} \qquad j=2,3,\ldots,J$$

and

$$\lambda_{01}=\lambda_{10}=0$$

The formulation of the model allowed site k to invade site j only when sites j and k were different colours. Now amend the model to allow site k to invade site j at rate λ_{jk} even when sites j and k are the same colour—these additional invasions will of course result in no change of any site colour. Thus as far as the colouring of the sites of the system is concerned the amendment will not affect the stochastic behaviour of the system. It will, however, mean that invasions of site j from site k form a Poisson process of rate λ_{jk} and that as j and k vary they index independent Poisson processes.

Suppose now that we know the exact moments within the interval $(0, t)$ at which site j is invaded from site k for $j, k=0, 1, \ldots, J$. From this information can we discover the colour of site j at time t? Consider the following method. Starting from time t look backwards through time to discover when site j was last invaded and from whence. If it was last invaded after time $t=0$ from another site of the set $\{2, 3, \ldots, J\}$, then look further back in time to discover when and from whence this site was last invaded, and so on. Remembering that the moments of invasion form realizations from independent Poisson processes it becomes clear that as we trace the origin of site j's colour backwards through time we will be following a random walk with transition intensities λ_{jk}, for $j, k=0, 1, \ldots, J$. If this random walk reaches site 1 within a time t then site j at time t is black; otherwise it is white. Thus $p_j(t)$ is simply the probability that a random walk starting at site j and with transition rates λ_{jk}, for $j, k=0, 1, \ldots, J$, reaches site 1 before a time t has elapsed. If the random walk on the set $\{2, 3, \ldots, J\}$ defined by the transition intensities λ_{jk}, for $j, k=2, 3, \ldots, J$, is reversible then we can as outlined in Section 5.1 obtain an electrical analogy. The random walk will be reversible if the λ_{jk} are symmetric. It will also be reversible if the graph defined on the set $\{2, 3, \ldots, J\}$ by the λ_{jk} (with an edge joining nodes j and k if either λ_{jk} or λ_{kj} is positive) is a tree (Lemma 1.5). Even when an electrical analogy exists it is not as useful as in the last section—there is nothing in the model which can be readily related to a flow of current.

It is worth emphasizing that the analysis of the invasion model does not rely upon the random walk defined by the transition intensities λ_{jk} being reversible. The origin of a site's colour can be traced backwards through time whatever the transition intensities. There is thus a contrast with the flow model of the previous section where the analysis breaks down unless the λ_{jk} are symmetric.

Exercises 5.3

1. In the analysis of the basic invasion model it was assumed that invasions of site j from site k form a Poisson process and that for distinct pairs (j, k) the Poisson processes are independent. Show that the analysis does not depend upon the independence assumption by considering the following cases:
   (i) Whenever site j invades, it simultaneously invades every site it is adjacent to.
   (ii) When site j invades site i, site l invades site k.
   In case (i) show that an electrical analogue exists in which all the resistors joining sites have the same resistance.
2. Generalize the invasion model to allow more than two colours.
3. Consider the following stochastic model of group decision making. A group consists of n individuals, and initially individual j holds opinion (or view) v_j, $j = 1, 2, \ldots, n$. If at time t individuals j and k hold differing opinions then the probability that individual j is convinced by individual k and changes his opinion to that of individual k in the interval $(t, t+\delta t)$ is $\lambda_{jk}\, \delta t + o(\delta t)$. Assume that between any two individuals there exists the possibility of communication, either directly or indirectly via a chain of other individuals. Show that the group will ultimately agree on view v_j with probability α_j where α_j is the solution to the equations

$$\alpha_j \sum_k \lambda_{jk} = \sum_k \alpha_k \lambda_{kj}$$

$$\sum_k \alpha_k = 1$$

4. In the preceding exercise suppose the probability that individual j concedes to individual k in the interval $(t, t+\delta t)$ is $\lambda_{jk} f(t)\, \delta t + o(\delta t)$; for example if $f(t)$ is a decreasing function then individuals become more stubborn as time progresses. Show that the conclusion remains the same provided $\int_0^\infty f(t)\, dt$ is infinite.
5. In Exercise 5.3.3 suppose the probability that individual j concedes to individual k in the interval $(t, t+\delta t)$ is $\lambda_{jk} f_{jk}(t)\, \delta t + o(\delta t)$ where

$$f_{jk}(t) = f_{kj}(t) \qquad \text{for} \quad t > 0, \quad j, k = 1, 2, \ldots, n$$

   These rates might arise if the degree of contact between individuals varies with time. Show that the conclusion remains the same provided $f_{jk}(t)$ is a bounded function of t, $\int_0^\infty f_{jk}(t)\, dt$ is infinite, and $\alpha_j \lambda_{jk} = \alpha_k \lambda_{kj}$ for $j, k = 1, 2, \ldots, n$. Deduce the corresponding result when $(f_{jk}(t); j, k = 1, 2, \ldots, n)$ is itself a stochastic process.

CHAPTER 6

Reversible Migration Processes

In this chapter we shall consider an adaptation of the migration processes of Chapter 2 which allows the probability intensity that a customer moves from one colony to another to depend upon the number in the receiving colony. The adapted process thus permits blocking, but in order to make any analytical progress reversibility must be assumed. In Section 6.2 we shall consider an application of the resulting process to the modelling of social grouping behaviour. Finally in Section 6.3 we shall contrast various processes, including those introduced in Chapters 2 and 5, in which at most one individual is allowed in each colony.

6.1 MIGRATION PROCESSES REVISITED

In Section 2.3 we considered a closed migration process with state given by $\mathbf{n}=(n_1, n_2, \ldots, n_J)$, where n_j is the number of individuals in colony j. If the probability intensity that an individual moves from colony j to k is

$$q(\mathbf{n}, T_{jk}\mathbf{n})=\lambda_{jk}\phi_j(n_j) \tag{6.1}$$

then we saw that the equilibrium distribution takes a simple form (Theorem 2.3). A limitation of the transition rate (6.1) is that it does not depend upon n_k, the number of individuals already present at colony k. Perhaps the simplest form incorporating such a dependence is

$$q(\mathbf{n}, T_{jk}\mathbf{n})=\lambda_{jk}\phi_j(n_j)\psi_k(n_k) \tag{6.2}$$

where $\phi_j(0)=0$ and for simplicity $\lambda_{jj}=0$. To ensure that $\mathbf{n}$ is irreducible within the state space

$$\mathcal{S}=\left\{\mathbf{n}\mid n_j\geq 0, j=1,2,\ldots,J; \sum_{j=1}^{J} n_j=N\right\}$$

we require that $\phi_j(n)>0$ if $n>0$, $\psi_j(n)>0$ if $n\geq 0$, and that the parameters λ_{jk} allow an individual to pass between any two colonies, either directly or indirectly via a chain of other colonies. With transition rates (6.2) the equilibrium equations become

$$\pi(\mathbf{n})\sum_{j=1}^{J}\sum_{k=1}^{J}\lambda_{jk}\phi_j(n_j)\psi_k(n_k)=\sum_{j=1}^{J}\sum_{k=1}^{J}\pi(T_{jk}\mathbf{n})\lambda_{kj}\phi_k(n_k+1)\psi_j(n_j-1)$$

In general these equations do not have a simple solution. We might hope, however, that we could solve the equations in the special case when the process $\mathbf{n}$ is reversible. The detailed balance conditions are then

$$\pi(\mathbf{n})\lambda_{jk}\phi_j(n_j)\psi_k(n_k) = \pi(T_{jk}\mathbf{n})\lambda_{kj}\phi_k(n_k+1)\psi_j(n_j-1)$$

and it is easy to check that a solution to these is

$$\pi(\mathbf{n}) = B\prod_{j=1}^{J}\left\{\alpha_j^{n_j}\prod_{r=1}^{n_j}\frac{\psi_j(r-1)}{\phi_j(r)}\right\} \tag{6.3}$$

provided

$$\alpha_j\lambda_{jk} = \alpha_k\lambda_{kj} \tag{6.4}$$

Thus although the transition rates (6.2) appear more general than (6.1) to deal with them we have to impose the restriction (6.4) on the parameters λ_{jk}. The normalizing constant B is, as usual, chosen so that the distribution $\pi(\mathbf{n})$ sums to unity over the state space $\mathcal{S}$. We can summarize the above results in the following theorem.

Theorem 6.1. *A stationary closed migration process with transition rates* (6.2) *is reversible if there exist positive constants* $\alpha_1, \alpha_2, \ldots, \alpha_J$ *satisfying condition* (6.4). *In this case the equilibrium distribution takes the form* (6.3).

For an open migration process we need to specify additional transition rates to complement (6.2). The most obvious choices are

$$q(\mathbf{n}, T_{j.}\mathbf{n}) = \mu_j\phi_j(n_j) \tag{6.5}$$

for the probability intensity that an individual leaves the system from colony j and

$$q(\mathbf{n}, T_{.k}\mathbf{n}) = \nu_k\psi_k(n_k) \tag{6.6}$$

for the probability intensity that an individual enters the system to join colony k. Assume that the parameters λ_{jk}, μ_j, and ν_k allow an individual to reach any colony from outside the system and to leave the system from any colony, either directly or indirectly via a chain of other colonies. This, together with the earlier assumption that $\phi_j(0)=0$, $\phi_j(n)>0$ if $n>0$, $\psi_j(n)>0$ if $n\geq 0$, ensures that the process $\mathbf{n}$ is irreducible within the state space $\mathbb{N}^J$. It is easy to check that the form (6.3) will again satisfy the detailed balance conditions provided

$$\alpha_j\mu_j = \nu_j \tag{6.7}$$

and condition (6.4) are satisfied. We thus have the following result.

Theorem 6.2. *A stationary open migration process with transition rates* (6.2), (6.5), *and* (6.6) *is reversible if there exist positive constants*

$\alpha_1, \alpha_2, \ldots, \alpha_J$ satisfying conditions (6.4) and (6.7). In this case the equilibrium distribution takes the form (6.3) and in equilibrium $n_1, n_2, \ldots, n_J$ are independent.

We shall call the processes introduced in this section reversible migration processes. This should not cause confusion with the migration processes of Chapter 2 since if a migration process as defined there is reversible then it is in fact a reversible migration process as defined here with $\psi_j(n)=1$ for $n \geq 0$, $j=1,2,\ldots,J$.

The results obtained in Chapter 3 show that various modifications can be made to a migration process without affecting the equilibrium distribution $\pi(\mathbf{n})$. In particular, suppose that when an individual arrives at colony j he is assigned a nominal lifetime which has an arbitrary distribution with unit mean, that he ages through this lifetime at rate $\phi_j(n_j)\sum_i \lambda_{ji}$ while there are n_j individuals in colony j, and that when his lifetime in colony j comes to an end he moves to colony k with probability $\lambda_{jk}/\sum_i \lambda_{ji}$. Suppose for simplicity that all nominal lifetimes are independent. The case where nominal lifetimes are exponentially distributed corresponds to a closed migration process with transition rates (6.1), but the equilibrium distribution $\pi(\mathbf{n})$ is the same whatever the nominal lifetime distributions, since the colonies are examples of server-sharing queues. Consider now a closed reversible migration process with transition rates (6.2). This can be modified to allow arbitrary nominal lifetimes by supposing that an individual in colony j ages through his lifetime at rate $\phi_j(n_j)\sum_i \lambda_{ji}\psi_i(n_i)$ and that when his lifetime in colony j comes to an end he moves to colony k with probability $\lambda_{jk}\psi_k(n_k)/\sum_i \lambda_{ji}\psi_i(n_i)$. Observe that when nominal lifetimes are exponentially distributed the process $\mathbf{n}$ is Markov with transition rates (6.2). In Chapter 9 (Exercises 9.3.2 and 9.4.2) we shall see that provided equations (6.4) have a solution the equilibrium distribution $\pi(\mathbf{n})$ is the same whatever the form of the nominal lifetime distributions.

Exercises 6.1

1. Use Kolmogorov's criteria to show that a stationary migration process with transition rates (6.2) is reversible only if condition (6.4) can be satisfied.
2. Consider a Markov process $\mathbf{n}$ for which the transition rates (6.5) and (6.6) are positive, and where these are not the only positive transition rates. Show that if $\mathbf{n}$ is reversible then $n_1, n_2, \ldots, n_J$ are independent and the equilibrium distribution is of the form (6.3), whatever the form of the transition rates other than (6.5) and (6.6).
3. If we regard the open migration process discussed in Section 2.4 as a model of the movement of particles between cells then the number of

particles in a cell will have a Poisson distribution if $\phi(n)=\phi n$, a geometric distribution if $\phi(n)=\phi$, $n>0$, and a Bernoulli distribution (i.e. a distribution on the values of zero or unity) if $\phi(1)=\phi$, $\phi(n)=\infty$, $n>1$. Physicists attach the names Maxwell–Boltzmann, Bose–Einstein, and Fermi–Dirac respectively to these distributions. Observe that the suggested functions $\phi(n)$ are the only ones which will give rise to these distributions. Discuss how the distributions could arise from the reversible migration processes of this section. Observe that there are a variety of models which could lead to each of the given distributions.

4. In an open migration process with transition rates (6.2), (6.5), and (6.6) the stream of individuals entering the system at colony j will not in general be Poisson. Suppose, however, that $\psi_j(n)\leq 1$ for $n\geq 0$, $j=1,2,\ldots,J$. In this case the given transition rates could be reconciled with Poisson streams if it is assumed that an individual arriving at colony j to find n_j individuals already there is lost with probability $1-\psi_j(n_j)$. If individuals arriving at and departing from colony j are considered as customers of class j show that, counting lost customers, the system is quasi-reversible under the weaker assumptions described in Exercise 3.2.1. The weaker assumptions are needed since the class of a customer may change as he passes through the system. Observe that in this system the probability that a customer is lost can depend upon the state of an individual colony in a way which could not be allowed in Section 3.5.
5. In an open migration process with transition rates (6.2), (6.5), and (6.6) suppose that if $\nu_k>0$ then $\psi_k(n)=1$, $n\geq 0$. Show that if individuals arriving from outside the system at colony j or leaving the system from colony j are considered as customers of class j then the system is quasi-reversible under the weaker assumptions described in Exercise 3.2.1. Of course, if all customers are considered to be of the same class the system is quasi-reversible, and thus Exercise 3.2.4 shows how to produce a quasi-reversible system in which the class of a customer does not change.

6.2 SOCIAL GROUPING BEHAVIOUR

One area where reversible migration processes can be useful is in the modelling of the behaviour of individuals gathering in groups for social reasons, e.g. monkeys forming sleeping groups or children at play. To illustrate the results of the previous section we shall consider an open and closed version of a model which might be appropriate in this context.

Suppose that group j consists of those individuals at a particular geographical location. Let n_j be the number of individuals in group j and suppose that $\mathbf{n}=(n_1, n_2, \ldots, n_J)$ is a migration process with the following transition rates:

$$q(\mathbf{n}, T_{\cdot k}\mathbf{n}) = a_k + c_k n_k$$
$$q(\mathbf{n}, T_{j\cdot}\mathbf{n}) = d_j n_j \qquad (6.8)$$
$$q(\mathbf{n}, T_{jk}\mathbf{n}) = \lambda_{jk} d_j n_j (a_k + c_k n_k)$$

where $\lambda_{jk} = \lambda_{kj}$, $a_k > 0$, and $d_k > c_k > 0$. Here a_k can be thought of as the attractiveness to an outsider of belonging to group k, c_k as the attractiveness to an outsider of an individual in group k, d_j as the propensity to depart from group j of an individual in group j, and λ_{jk} as a measure of the proximity of groups j and k. The transition rates (6.8) are of the form discussed in the previous section with

$$\phi(n_j) = d_j n_j \qquad \text{and} \qquad \psi_k(n_k) = a_k + c_k n_k$$

and a solution to equations (6.4) and (6.7) is $\alpha_1 = \alpha_2 = \cdots = \alpha_J = 1$. Thus Theorem 6.2 gives the form of the equilibrium distribution. To calculate the normalizing constant is not difficult, and the conclusion is that in equilibrium $n_1, n_2, \ldots, n_J$ are independent, each with a negative binomial distribution:

$$\pi(n_j) = \binom{f_j + n_j - 1}{n_j} (1 - g_j)^{f_j} g_j^{n_j} \qquad n_j = 0, 1, \ldots$$

where $f_j = a_j/c_j$ and $g_j = c_j/d_j$. The expected number of individuals in group j is thus $a_j/(d_j - c_j)$.

Consider now the closed version of the above model with transition rates

$$q(\mathbf{n}, T_{jk}\mathbf{n}) = \lambda_{jk} d_j n_j (a_k + c_k n_k) \qquad (6.9)$$

where $\lambda_{jk} = \lambda_{kj}$, a_k, c_k, $d_k > 0$, and the total number of individuals in the system is N. Theorem 6.1 allows us to deduce the form of the equilibrium distribution but the normalizing constant is in general an awkward expression. It simplifies when $c_j/d_j = g$ for $j = 1, 2, \ldots, J$. Then the equilibrium distribution can be written

$$\pi(\mathbf{n}) = \binom{-\sum f_k}{N}^{-1} \prod_{j=1}^{J} \binom{-f_j}{n_j} \qquad (6.10)$$

for $\mathbf{n}$ such that $\sum n_j = N$, where $f_j = a_j/c_j$.

A drawback of the model described is that it assumes a group is based at one of J geographical locations. Often this assumption is inappropriate, and in Chapter 8 we shall consider models which do not restrict the groups in this way.

Exercises 6.2

1. Show that if $c_k = 0$ in transition rates (6.8) then in equilibrium n_k has a Poisson distribution. Deduce that if $c_k = 0$ in transition rates (6.9) then in equilibrium $\mathbf{n}$ has a multinomial distribution.

2. Deduce from the distribution (6.10) that the marginal distribution for n_j is the Polya distribution

$$\pi(n_j) = \binom{-\sum f_k}{N}^{-1} \binom{-f_j}{n_j} \binom{f_j - \sum f_k}{N - n_j}$$

Suppose now that $f_1 = f_2 = \cdots = f_J$. Show that if $N, J \to \infty$ with N/J held constant the marginal distribution for n_j approaches a negative binomial distribution.

3. In the open model with transition rates (6.8) determine the probability flux that an individual moves from group j to group k. Use Little's result to deduce the mean time an individual stays in group j.

6.3 CONTRASTING FLOW MODELS

Section 5.2 discussed a flow model in which each site could hold at most one individual. This restriction can also be imposed on the migration processes of Chapter 2 or of this chapter, with rather different effects. In this section we shall comment briefly on the resulting flow models and introduce a further one.

Consider an open migration process with transition rates (2.8) in which

$$\phi(n) = \begin{cases} 1 & n = 1 \\ \infty & n > 1 \end{cases}$$

In this process site j can hold at most one individual. If an individual arrives at site j when it is already occupied he is immediately ejected from the site and moves on to site k with probability λ_{jk}/λ_j or leaves the system with probability μ_j/λ_j. Thus there is no blocking in this system—the reverse in fact, since the more likely site j is to be occupied the faster an individual needing to visit site j will pass through the system. If $\alpha_1, \alpha_2, \ldots, \alpha_J$ is the solution to equations (2.9) then in equilibrium site j is occupied with probability

$$\frac{\alpha_j}{1 + \alpha_j} \tag{6.11}$$

independently of the state of the rest of the system.

Consider now an open migration process with transition rates given by (6.2), (6.5), and (6.6) where

$$\phi(n) = 1 \qquad n > 0$$

$$\psi(n) = \begin{cases} 1 & n = 0 \\ 0 & n > 0 \end{cases}$$

In this process site j can again hold at most one individual. While site j is occupied transitions which would bring another individual to site j are forbidden. Provided conditions (6.4) and (6.7) are satisfied we can determine the equilibrium distribution, and in equilibrium site j is again occupied with probability (6.11), independently of the rest of the system. A drawback of the model is that since it is reversible there can be no net flow of individuals in a given direction through the system. Note that in both this model and in the flow model of Section 5.2 an individual unable to move to site j from site k because site j is occupied may well end up moving from k to a site other than j.

There is a further flow model for which analytical results are available, and a closed version of it can be described as follows. Suppose that while site j is occupied and site k is free the probability intensity that the individual at j moves to k is λ_{jk} $(j, k=1, 2, \ldots, J)$ where the parameters λ_{jk} satisfy

$$\sum_k \lambda_{jk} = \sum_k \lambda_{kj} \qquad j=1, 2, \ldots, J \tag{6.12}$$

The equilibrium equations for this process are

$$\pi(\mathbf{n}) \sum_{j:n_j=1} \sum_{k:n_k=0} \lambda_{jk} = \sum_{j:n_j=1} \sum_{k:n_k=0} \pi(T_{jk}\mathbf{n})\lambda_{kj} \tag{6.13}$$

A solution to these equations is (Exercise 6.3.3)

$$\pi(\mathbf{n}) = \binom{J}{N}^{-1} \tag{6.14}$$

for each $\mathbf{n}$ representing a state where there are N individuals present in the system, all at different sites. Thus all possible states are equally likely.

An open version of the above model can be obtained by appending to the system a large number of additional sites, each connected to the previously existing sites in the same way. The details are given in Exercise 6.3.5 and the resulting model can be described as follows. If site j is occupied then the individual there leaves the system with probability intensity μ_j; if site k is free an individual arrives there from outside the system with probability intensity ν_k; and if site j is occupied and site k is free then the individual at j moves to k with probability intensity λ_{jk}. In place of restriction (6.12) assume that the rates satisfy the equations

$$\sum_k \lambda_{jk} + \frac{\mu_j}{1-p} = \sum_k \lambda_{kj} + \frac{\nu_j}{p} \qquad j=1, 2, \ldots, J \tag{6.15}$$

where

$$p = \frac{\rho}{1+\rho} \tag{6.16}$$

and

$$\rho = \frac{\sum_j \nu_j}{\sum_j \mu_j} \tag{6.17}$$

The equilibrium equations for the process are

$$\pi(\mathbf{n})\left[\sum_{j:n_j=1}\sum_{k:n_k=0}\lambda_{jk} + \sum_{j:n_j=1}\mu_j + \sum_{k:n_k=0}\nu_k\right]$$

$$= \sum_{j:n_j=1}\sum_{k:n_k=0}\pi(T_{jk}\mathbf{n})\lambda_{kj} + \sum_{j:n_j=1}\pi(T_{j\cdot}\mathbf{n})\nu_j + \sum_{k:n_k=0}\pi(T_{\cdot k}\mathbf{n})\mu_k$$

The equilibrium equations are satisfied by

$$\pi(\mathbf{n}) = B\rho^{\sum_j n_j} \tag{6.18}$$

and hence in equilibrium a site is occupied with probability p, and whether the site is occupied or not is independent of the state of the rest of the system. This is an intriguing result: we are accustomed to such independence in open networks of quasi-reversible queues and in reversible migration processes, but this flow model shows that it can occur in other systems as well.

As an example of the result consider the one-dimensional flow model illustrated in Fig. 6.1, where jumps take place between adjacent sites with the probability intensities shown. If

$$\lambda_1 = \lambda_2 + \mu + \nu$$

then restriction (6.15) is satisfied and in equilibrium a site is occupied with probability $\nu/(\mu+\nu)$, independently of the other sites. The same model was considered in Exercise 5.2.2 under the restriction $\lambda_1 = \lambda_2$.

Exercises 6.3

1. Observe that in the first flow model considered in this section the probability that site j is occupied, given by expression (6.11), remains unaltered when the time a particle remains at site j is arbitrarily distributed with mean λ_j^{-1}. In the case where the distribution is exponential observe that the model is unaltered if it is the previous occupant of site j who is expelled when a second individual arrives there.
2. The organizational hierarchy illustrated in Fig. 6.2 consists of J posts, each of which is held by at most one individual. At points in time which

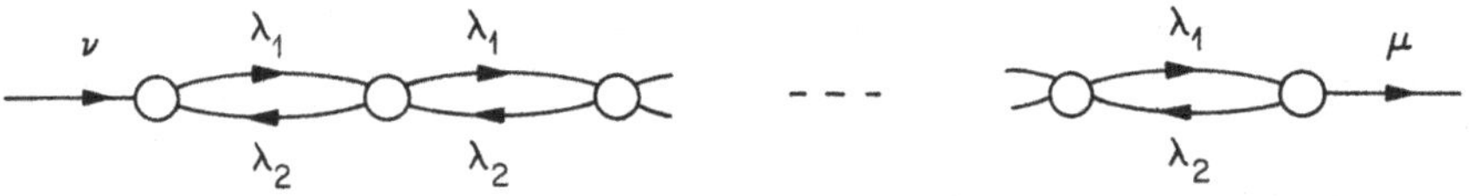

Fig. 6.1 A one-dimensional flow model

Fig. 6.2 An organizational hierarchy

form a Poisson process of rate ν the most senior individual in the organization leaves (to join the Head Office). This causes promotions within the organization in the following way. When post j becomes vacant there is a delay, exponentially distributed with mean λ_j^{-1}, before a replacement takes up the post. The replacement comes from the next most junior post in the organization which is occupied, or from outside the organization if all the more junior posts are vacant. By considering the movement of vacancies show that in equilibrium the posts are vacant or not independently and the probability post j is vacant is $\nu/(\nu+\lambda_j)$. Show that this remains true if the delay before a replacement is appointed to post j is arbitrarily distributed with mean λ_j^{-1}.

Extend the model to allow K_j individuals to hold positions at level j. Show that the probability level j has its complete complement of K_j individuals is

$$\left[\sum_{n=0}^{K_j} \frac{1}{n!}\left(\frac{\nu}{\lambda_j}\right)^n\right]^{-1}$$

3. Consider a Markov process with states $1, 2, \ldots, J$ and transition rates satisfying equation (6.12). Observe that in equilibrium each state is equally likely. Deduce from Lemma 1.4 that

$$\sum_{j\in\mathcal{A}} \sum_{k\notin\mathcal{A}} \lambda_{jk} = \sum_{j\in\mathcal{A}} \sum_{k\notin\mathcal{A}} \lambda_{kj}$$

and hence that expression (6.14) satisfies equations (6.13).

4. Consider the model of a mining operation discussed in Section 2.3. Suppose now that a machine cannot start work on a face until the next face is free. Show that machine j operates as a queue with $\phi_j(1)=0$, $\phi_j(n)=\phi_j$, $n>1$. Deduce that in equilibrium the probability machine j is working is

$$\frac{B_{N-J}}{\phi_j B_{N-J-1}}$$

where B_N is as defined for the original model of Section 2.3. Observe that if $\phi_1=\phi_2=\cdots=\phi_J=\phi$ the system can be represented by a flow model satisfying condition (6.12). Show that in this case the average time for a machine to complete one cycle of faces is

$$\frac{N(N-1)}{(N-J)\phi}$$

An alternative amendment to the model would be to suppose that after a machine has finished work on a face it cannot move on until the next face is free. None of the flow models considered can represent this behaviour, and indeed analytical results are generally unavailable for networks involving this fairly common form of blocking.

5. An open flow model satisfying (6.15) can be obtained as the limiting case of a closed flow model formed as follows. The closed flow model consists of the given J sites together with M appended sites. The flow rate from site j to each appended site is $\mu_j/M(1-p)$, and from each appended site to site j is ν_j/Mp. Show that for this closed flow model condition (6.12) becomes (6.15) provided p satisfies (6.16) and (6.17). Now let the number of appended sites M and the number of individuals N tend to infinity in such a way that N/M tends to p. Show that in the limit the open flow model leading to distribution (6.18) is obtained.
6. Using the transition rates (5.14), (5.15), and (5.16) show how the flow models leading to the equilibrium distributions (6.14) and (6.18) can be extended to allow a site to contain more than one individual. Observe that in equilibrium the number of individuals at a site in the open version will be binomially distributed.
7. Let $\mathbf{n}(t)$ be the state at time t of the flow model leading to equation (6.14). Show that the reversed process $\mathbf{n}(-t)$ corresponds to a similar flow model, but with λ_{jk} replaced by λ_{kj}. If $\mathbf{n}(t)$ is the state at time t of the open flow model leading to equation (6.18) show that $\mathbf{n}(-t)$ corresponds to a similar flow model with λ_{jk}, ν_j, and μ_j replaced by λ_{kj}, $\rho\mu_j$, and ν_j/ρ respectively. Deduce that the one-dimensional flow model considered at the end of this section is dynamically reversible.
8. A further example of an open flow model satisfying (6.15) is given by the following choice of parameters:

$$\lambda_{jk}=0 \qquad \text{unless } k=j+1$$

$$\lambda_{j,j+1}=\lambda$$

for $j=1,2,\dots,J-1$

$$\lambda_{Jk}=0 \qquad \text{unless } k=1$$

$$\lambda_{J1}=\lambda$$

and

$$\left.\begin{aligned}\mu_j&=\mu\\ \nu_j&=\nu\end{aligned}\right\} \qquad \text{for } j=1,2,\dots,J$$

Check that in equilibrium the probability that a given site is occupied is ν/μ and hence does not depend upon λ. Show that the process is dynamically reversible. Following Exercise 6.1.4 discuss how the system might be rendered quasi-reversible.

CHAPTER 7
Population Genetics Models

One theory of evolution holds that favourable mutations are relatively rare while in contrast selectively neutral mutations are common and account for much of the diversity between individuals observed at the molecular level. In this chapter a stochastic model is discussed which provides some insight into the behaviour of a population subject to recurrent neutral mutation. The model, introduced below, is closely related to the migration processes of Chapters 2 and 6 and the invasion model of Chapter 5, but the major motivation for its inclusion is the use made of reversibility in Section 7.2 to elucidate some of its properties.

7.1 NEUTRAL ALLELE MODELS

Consider a population of M individuals in which the individuals are of various genetic (or allelic) types. Suppose there are J types (or alleles) altogether, and let n_j be the number of individuals of allelic type j. The mechanism by which the population reproduces is as follows. Individuals die at rate μ. When a death occurs an individual, chosen at random from amongst the remaining $M-1$ individuals, gives birth. The offspring is of the same allelic type as the parent with probability $1-u$ and is a mutation of a different allelic type with probability u. When a mutation occurs the mutant individual is equally likely to have any of the other $J-1$ allelic types, excluding his parent's type. In this model the population size remains constant at M and the alleles are neutral, in the sense that an individual's type does not affect his ability to survive or to produce offspring.

It follows from the above description that the Markov process $\mathbf{n} = (n_1, n_2, \ldots, n_J)$ has transition rates

$$q(\mathbf{n}, T_{jk}\mathbf{n}) = \mu \frac{n_j}{M} \frac{n_k(1-u)+(M-n_k-1)u/(J-1)}{M-1}$$

These are of the form (6.2), and hence the equilibrium distribution is given by Theorem 6.1. Indeed these transition rates are a special case of the form (6.9), and so from expression (6.10) it follows that the equilibrium distribution is

$$\pi(\mathbf{n}) = \binom{-Jf}{M}^{-1} \prod_{j=1}^{J} \binom{-f}{n_j} \tag{7.1}$$

where

$$f = \frac{(M-1)u}{J(1-u)-1} \tag{7.2}$$

The above model is of more interest when there are an infinite number of alleles, so that when a mutation occurs the mutant individual has a completely new allelic type, never before represented in the population. Letting $J \to \infty$ in expression (7.1) causes problems: the probability that any *given* allele is present in the population will tend to zero. It is more helpful to describe the population by the process $\mathbf{M} = (M_1, M_2, \ldots, M_M)$ where M_i is the number of alleles represented in the population by i individuals. Thus

$$\sum_{i=1}^{M} iM_i = M \tag{7.3}$$

It follows from the distribution (7.1) that the equilibrium distribution for the process $\mathbf{M}$ is

$$\pi_M(\mathbf{M}) = \frac{J!}{M_1!\,M_2!\cdots M_M!\,(J-\sum M_i)!}\binom{-Jf}{M}^{-1}\binom{-f}{1}^{M_1}\binom{-f}{2}^{M_2}\cdots\binom{-f}{M}^{M_M} \tag{7.4}$$

Now let $J \to \infty$ with u held constant; then from (7.2)

$$Jf \to \frac{(M-1)u}{1-u} = \nu \tag{7.5}$$

say. In the limit the form (7.4) becomes (Exercise 7.1.3)

$$\pi_M(\mathbf{M}) = \binom{\nu+M-1}{M}^{-1}\prod_{i=1}^{M}\left(\frac{\nu}{i}\right)^{M_i}\frac{1}{M_i!} \tag{7.6}$$

for $\mathbf{M}$ satisfying (7.3). We shall call the resulting process $\mathbf{M}$ the *infinite alleles model.*

The distribution (7.6) is strikingly similar to the equilibrium distribution for the family size process discussed in Section 2.4, and we shall now show that the relationship is not coincidental. Consider a family size process in which individuals with a new allelic type join the population at rate $\lambda\nu$, individuals give birth to new individuals of the same allelic type at rate λ, and individuals die at rate μ (this is a slight change from the process considered in Section 2.4: we have replaced ν by $\lambda\nu$). It follows from the discussion contained in Section 2.4 that if M_i is the number of alleles represented in the population by i individuals then $\mathbf{M} = (M_1, M_2, \ldots)$ is reversible with equilibrium distribution

$$\pi(\mathbf{M}) = \prod_{i=1}^{\infty} e^{-\alpha_i}\frac{\alpha_i^{M_i}}{M_i!}$$

where

$$\alpha_i = \frac{\nu}{i}\left(\frac{\lambda}{\mu}\right)^i$$

provided $\lambda < \mu$. Thus $M_1, M_2, \ldots$ are independent, each with a Poisson distribution. Now suppose that we truncate the process $\mathbf{M}$ by forbidding transitions which would cause the total number of individuals alive to drop below $M-1$ or rise above M. Then Corollary 1.10 shows that the equilibrium distribution will have the form

$$\pi(\mathbf{M}) = B \prod_{i=1}^{M} \frac{\alpha_i^{M_i}}{M_i!} \tag{7.7}$$

for $\mathbf{M}$ such that $\sum iM_i = M-1$ or M. How does the process $\mathbf{M}$ behave when its state space is truncated in this way? Well, when the population size is M any particular existing individual dies with probability intensity μ. When the population size is $M-1$ any particular existing individual gives birth to another individual of the same type with probability intensity λ, and with probability intensity $\lambda\nu$ a mutant individual of a new allelic type is born. The proportion of individuals born which are mutations is $\nu/(\nu+M-1)$, which equals u, by relation (7.5). When the population size is $M-1$ the process thus behaves as if each existing individual gives birth at rate $(\nu+M-1)\lambda$, and with probability u the individual born is a mutation. If we now let $\lambda \to \infty$ the births occur immediately after deaths and we obtain the infinite alleles model which led to the distribution (7.6). Distribution (7.7) can be rewritten

$$\pi(\mathbf{M}) = B\left(\frac{\lambda}{\mu}\right)^{\sum iM_i} \prod_{i=1}^{M} \left(\frac{\nu}{i}\right)^{M_i} \frac{1}{M_i!}$$

for $\mathbf{M}$ such that $\sum iM_i = M-1$ or M, and as $\lambda \to \infty$ this approaches the distribution (7.6), as of course it must do.

Often it is not possible to observe the entire population, but only a sample from it. We shall conclude this section by obtaining the sampling distribution for a sample from the infinite alleles model. Say that a set of M individuals has the *description* $\mathbf{M} = (M_1, M_2, \ldots)$ if there are M_i alleles represented by i individuals in the set. The following theorem shows that the equilibrium distribution (7.6) of the infinite alleles model has a rather interesting property.

Theorem 7.1. *Suppose that a random sample of size m is chosen without replacement from a population of size M, $m \leq M$. If the population has the description $\mathbf{M}$ with probability $\pi_M(\mathbf{M})$ then the sample has the description $\mathbf{m}$ with probability $\pi_m(\mathbf{m})$.*

Proof. Consider the process **M** described above in which the population size fluctuates between $M-1$ and M. The equilibrium distribution is given by (7.7) and hence the description of the population conditional on the population size being $M-1$ is $\pi_{M-1}(\mathbf{M})$, and conditional on it being M is $\pi_M(\mathbf{M})$. Now when the population size drops from M to $M-1$ the effect is the same as that of choosing a random sample of size $M-1$. Thus when a random sample of size $M-1$ is chosen from a population whose description is **M** with probability $\pi_M(\mathbf{M})$, the description of the sample must be **m** with probability $\pi_{M-1}(\mathbf{m})$. This establishes the theorem for the case $m=M-1$. For general m the theorem follows from the observation that one way to choose a sample of size m is first to choose a sample of size $M-1$, then from this to choose a sample of size $M-2$, and so on until only m individuals are included.

Theorem 7.1 shows that the sampling distribution for a sample of size m from a population of size M is the same as the equilibrium distribution for a population of size m, provided both populations have the same value of ν.

Exercises 7.1

1. Show that the model described at the beginning of this section remains a reversible migration process if the mean lifetime an individual of allelic type j is μ_j^{-1} and if the offspring of an individual of type j is of type k with probability $up_k(1-p_j)$, where $\sum p_j = 1$.
2. A population of size M is divided into J types in accordance with expression (7.1), and a random sample of size m ($\leq M$) is chosen from it. Write down the conditional distribution for the composition of the sample given the composition of the population. Deduce that the sample will divide into J types in accordance with expression (7.1), but with M replaced by m.
3. By considering the coefficients of x^M in the power series expansions of the identity

$$(1-x)^{-\nu} = \prod_{j=1}^{\infty} e^{\nu x^j/j}$$

show that the distribution (7.6) sums to unity. Deduce (7.6) from (7.4).
4. Suppose that **M** is distributed according to expression (7.6). Show that if a random sample of size m has the description **m**, with a particular allele represented by i individuals in the sample, then an individual randomly selected from the remaining $M-m$ members of the population is of that particular allelic type with probability $i/(m+\nu)$.
5. The heterozygosity of a population is defined to be the probability that two distinct individuals, chosen at random from the population, are of different allelic types. Deduce from Theorem 7.1 that the heterozygosity

of a population is $\nu/(\nu+1)$ and that the whole population is of the same allelic type with probability

$$\binom{\nu+M-1}{M-1}^{-1}$$

6. If $\pi_M(\mathbf{M})$ is given by expression (7.6) then Theorem 7.1 has established the following.
   (a) If a random sample of size m is chosen from a set of size M whose description is $\mathbf{M}$ with probability $\pi_M(\mathbf{M})$, then the description of the random sample is $\mathbf{m}$ with probability $\pi_m(\mathbf{m})$.

   Show also that

   (b) If an individual is chosen at random from a set of size M whose description is $\mathbf{M}$ with probability $\pi_M(\mathbf{M})$ and if the individual is found to be of the same allelic type as exactly $M-m-1$ others in the set then the remaining m individuals form a set whose description is $\mathbf{m}$ with probability $\pi_m(\mathbf{m})$.
7. (*Hard*) The preceding exercise has a converse. Suppose that $\pi_M(\mathbf{M})$ is a probability distribution over descriptions of a set of size M and that $\pi_M(\mathbf{M})>0$ for all $\mathbf{M}$ satisfying equation (7.3). Show that if statements (a) and (b) hold for all m and M satisfying $m<M$ then $\pi_M(\mathbf{M})$ must be of the form (7.6).
8. If $\mathbf{M}$ is distributed according to (7.6) with ν an unknown parameter show that the number of alleles in the population, $\sum M_i$, is a sufficient statistic for ν. If only a random sample from the population is observed deduce that the number of alleles present in the sample is sufficient for ν.
9. In the models described in this section it has been assumed that individuals' lifetimes are exponentially distributed with mean μ^{-1}. In fact the equilibrium distributions (7.1) and (7.6) remain unaltered if individuals' lifetimes are arbitrarily distributed. Establish this by considering an infinite server queue at which arrival rates are of the form (3.27) with

$$\Psi(\mathbf{n})=\psi\left(\sum_{j=1}^{J} n_j\right)\prod_{j=1}^{J}(f+n_j-1)!$$

   If lifetimes are exponentially distributed then the number of offspring an individual has is geometrically distributed. If lifetimes are constant show that the number of offspring an individual has is binomially distributed.
10. The infinite alleles model and the family size process can be regarded as special cases of a more general stochastic population model. Let M_i be

the number of families of size i and let $M=\sum iM_i$ be the total population size. Suppose that $\mathbf{M}=(M_1, M_2, \ldots)$ is a Markov process with transition rates

$$\begin{aligned} q(\mathbf{M}, T_{i,i+1}\mathbf{M}) &= iM_i\lambda(M) \qquad i=1,2,\ldots \\ q(\mathbf{M}, T_{i,i-1}\mathbf{M}) &= iM_i\mu(M) \qquad i=2,3,\ldots \\ q(\mathbf{M}, T_{\cdot 1}\mathbf{M}) &= \nu\lambda(M) \\ q(\mathbf{M}, T_{1\cdot}\mathbf{M}) &= M_1\mu(M) \end{aligned}$$

Thus birth, death, and immigration rates are affected by the total population size. Verify that in equilibrium the process $\mathbf{M}$ is reversible with

$$\pi(\mathbf{M}) = B\left\{\prod_{r=1}^{M} \frac{\lambda(r-1)}{\mu(r)}\right\} \prod_{i=1}^{\infty} \left(\frac{\nu}{i}\right)^{M_i} \frac{1}{M_i!}$$

Show that if a random sample of size m is taken from the population then its description $\mathbf{m}$ has distribution (7.6) with M replaced by m. Show that this property of a random sample remains true if the population size at time $t=0$ is zero and the functions $\lambda(M)$ and $\mu(M)$ are replaced by functions of time $\lambda(M, t)$ and $\mu(M, t)$. Deduce that the property remains true if $\lambda(M, t)$ and $\mu(M, t)$ are themselves stochastic processes.

11. Suppose the family size process (or more generally the process described in the previous exercise) is used to model the numbers of moths present in a particular location, with each family representing a distinct species of moth. Deduce from Exercise 7.1.4 that if a trap catches a random sample of m moths then the expected number of species caught is

$$\sum_{i=1}^{m} \frac{\nu}{\nu+i-1}$$

Observe that this depends upon the parameter ν alone, unlike the corresponding relationship based upon the expected value of m, found in Exercise 2.4.7.

12. Consider a population process in which the population size fluctuates between M and M' as follows. After the population size has been M for a certain time a reproduction period is entered, when only births and immigrations are allowed. During this reproduction period each individual gives birth at rate λ to a new member of its family, and immigrants arrive at rate $\nu\lambda$ to found new families. Individuals appearing during the reproduction period are allowed to give birth themselves during that same period, and the period ends when the population size reaches M'. Then a random sample of M individuals is chosen from the

population to form the next generation and the procedure is repeated. Show that if the process **M** is observed in discrete time just before each reproduction period begins then it is reversible with equilibrium distribution (7.6). Show that the same is true if the process is observed immediately *after* each reproduction period, with M replaced by M' in expression (7.6).

13. Amend the model described in the previous exercise so that the reproduction period ends when the population size reaches $2M$, and then suppose the original M individuals die, leaving a new generation of size M. Show that the process **M** observed just before each reproduction period begins is identical to that for the unamended model with M' infinite.
14. Suppose that in the model of the previous exercise individuals appearing during a reproduction period are not allowed to give birth. The resulting model can be viewed as follows. The parent of each individual in the new generation is chosen independently and at random from among the M individuals alive in the previous generation, and each birth may result in a new mutation, mutations arising independently over the M births. Observe that the number of offspring an individual has is binomially distributed and that when an allele appears for the first time it is represented by just one individual in the population. Deduce that the process **M** obtained from this model is not reversible.

7.2 THE AGE OF AN ALLELE

Suppose that a population has been evolving according to the infinite alleles model for a long period. It is observed at time t, and it is found that K alleles are present in the population, M_i of them represented by i individuals, for $i = 1, 2, \ldots$. What does this information tell us about the ages of the K alleles?

We would like to be able to deduce from the reversibility of the process $\mathbf{M}(t)$ that the future of an allele represented in the population is stochastically similar to its past. Unfortunately such a conclusion does not immediately follow, since from a realization of the process $\mathbf{M}(t)$, $-\infty < t < +\infty$, it is not possible to discern the progress between first occurrence and eventual extinction of a particular allele. The problem could be overcome by using the finite allele model of the previous section and a limiting argument, but we shall use an alternative labelling method.

Suppose that the allelic type of each individual in the population is associated with an integer in the range from zero to M. When a non-mutant individual is born its allelic type is that of its parent. When a mutant individual is born its allelic type is not that of its parent, nor that of any of the other alleles present in the population. Since the population size is M

there will be at least one integer in the range from zero to M which can be assigned to the new allele; if there is more than one possible integer the choice is made at random. The integer associated with an allele is not intended to represent any physical characteristic of the allele—it simply labels it. As time progresses the same label will be used repeatedly for different alleles, but note that after an allele becomes extinct an interval will elapse before its label is used again. Describe the state of the population by

$$(\mathbf{M}, \mathbf{l}) = (M_1, M_2, \ldots, M_M; l(1,1), l(1,2), \ldots, l(1, M_1);$$
$$l(2,1), l(2,2), \ldots, l(2, M_2);$$
$$\ldots, l(M, M_M))$$

with $l(i, k)$ the label of the kth allele among those represented in the population by i individuals. The effect of the death of an individual whose allelic type had j representatives is to be equivalent to randomly choosing one of the M_j labels from among $l(j,1), l(j,2), \ldots, l(j, M_j)$ and inserting it at random into one of the $M_{j-1}+1$ positions among $l(j-1,1)$, $l(j-1,2), \ldots, l(j-1, M_{j-1})$. A birth is to be dealt with similarly.

Theorem 7.2. *The labelled process* $(\mathbf{M}, \mathbf{l})$ *is reversible and has equilibrium distribution*

$$\pi(\mathbf{M}, \mathbf{l}) = \frac{(M - \sum M_i + 1)!}{(M+1)!} \binom{\nu + M - 1}{M}^{-1} \prod_{i=1}^{M} \left(\frac{\nu}{i}\right)^{M_i} \frac{1}{M_i!}$$

Proof. The detailed balance conditions are easily checked and the result follows from these. Observe that the first term of the equilibrium distribution is the reciprocal of the number of distinct orderings of $\sum M_i$ different labels. In equilibrium $\mathbf{M}$ has the distribution (7.6) and given $\mathbf{M}$ every possible arrangement of labels, $\mathbf{l}$, is equally likely.

A realization of the labelled process $(\mathbf{M}, \mathbf{l})$ for $-\infty < t < \infty$ allows us to trace the history of any particular allele from the point in time when it first appears until it becomes extinct. Consider the problem of estimating, from an observation $\mathbf{M}$ on the process at a particular time t, the age of an allele present in the population. From the reversibility of the process $(\mathbf{M}, \mathbf{l})$ we see that the age of the allele has the same distribution as the time to extinction of that allele. But to calculate the distribution of the time to extinction it is not necessary to have the frequencies of the other alleles in the population: the future frequency of an allele in the population at time t follows a random walk with transition intensities

$$q(j, j-1) = \mu \frac{j}{M}\left(\frac{M-j}{M-1} + \frac{j-1}{M-1} u\right)$$
$$q(j, j+1) = \mu \frac{M-j}{M} \frac{j}{M-1} (1-u) \tag{7.8}$$

Thus the age distribution of an allele represented in the population by j individuals can be calculated (Exercise 7.2.1); it depends upon j but not upon the entire state $\mathbf{M}$.

The rest of this section will be devoted to some other consequences of Theorem 7.2, but before we embark on these it is worth clarifying the contribution reversibility makes to the solution of problems concerning the age of an allele. The labelled process $(\mathbf{M}, \mathbf{I})$ arising from some genetic models (e.g. that described in Exercise 7.1.14) is not reversible. Nevertheless, if it is a stationary Markov process then the reversed process obtained from it will also be a stationary Markov process, and given the state $(\mathbf{M}, \mathbf{I})$ at a fixed point in time the age of an allele in the original process will have the same distribution as the time to extinction of that allele in the reversed process. The difficulty is that the reversed process is likely to have extremely complex transition rates. The results obtained in this section follow from the tractability of the reversed process in the case where $(\mathbf{M}, \mathbf{I})$ is reversible.

Corollary 7.3. *If at a given time an allele is represented by j individuals in the population the probability that this allele is the oldest of the alleles then existing is j/M.*

Proof. Suppose that at a given time M different alleles are represented in the population; thus no two individuals have the same allelic type. The probability that a particular allele (*not* individual) will outlive the other $M-1$ alleles is M^{-1}, by symmetry. It follows from this that if at a given time an allele is represented by j individuals in the population the probability that this allele will outlive the other alleles then existing is j/M. The reversibility established in Theorem 7.2 implies that the probability this allele is the oldest is also j/M.

Corollary 7.4. *At a given time the age of the oldest allele is independent of its frequency in the population.*

Proof. The time to extinction of all the alleles currently existing in the population does not depend upon $\mathbf{M}$; it is simply the time a random walk with transition rates (7.8) takes to reach zero starting from M. The distribution of this time will remain unchanged if we are given which of the currently existing individuals is the ancestor of the last surviving individual with a currently existing allelic type, and so will remain unchanged if we are given which allele will survive the longest. Interchanging times to extinction with ages, as Theorem 7.2 allows us to do, gives the result.

An extension of Corollary 7.4 is given in Exercise 7.2.2.

Corollary 7.5. *In equilibrium the probability that the oldest allele existing is represented by i individuals in the population is, for $i = 1, 2, \ldots, M$,*

$$\frac{\nu}{M}\binom{\nu+M-1}{i}^{-1}\binom{M}{i} \tag{7.9}$$

Proof. Let the random variable x be the frequency of the oldest allele. Theorem 7.2 implies that x has the same distribution as the frequency of the allele which will survive the longest, but this in turn has the same distribution as the frequency of the allelic type of a randomly chosen individual from the population. Thus, using equation (7.6),

$$\begin{aligned}\text{Prob}\{x=i\} &= \sum_{\mathbf{M}} \text{Prob}\{x=i \mid \mathbf{M}\}\pi_M(\mathbf{M}) \\ &= \frac{\nu}{M}\binom{\nu+M-1}{M}^{-1}\binom{\nu+M-i-1}{M-i} \\ &= \frac{\nu}{M}\binom{\nu+M-1}{i}^{-1}\binom{M}{i}\end{aligned}$$

Suppose a sample of size m ($\leq M$) is chosen from the population. We have seen in Theorem 7.1 that the sample has the description $\mathbf{m}$ with probability $\pi_m(\mathbf{m})$. Given the description $\mathbf{m}$ of the sample it is possible to make some deductions about the relative ages of the alleles represented in the sample.

Theorem 7.6. *An allele represented by i individuals in a sample of size m is the oldest allele in the population with probability*

$$\frac{i(\nu+M)}{M(\nu+m)}$$

Proof. We must calculate the probability that an individual chosen at random from the population is of the given allelic type. With probability m/M the randomly chosen individual will belong to the sample, and if this happens the probability that it is of the given allelic type is i/m. With probability $(M-m)/M$ the randomly chosen individual will be in the $M-m$ members of the population outside the sample, and if this happens the probability that it is of the given allelic type is $i/(m+\nu)$ (Exercise 7.1.4). Thus the probability we are seeking is

$$\frac{m}{M}\frac{i}{m}+\frac{M-m}{M}\frac{i}{m+\nu}=\frac{i(\nu+M)}{M(\nu+m)}$$

Corollary 7.7. *The probability a sample of size m contains the oldest allele in the population is*

$$\frac{m(\nu+M)}{M(\nu+m)}$$

Proof. This follows directly from Theorem 7.6 by summing over all the alleles represented in the sample.

Corollary 7.8. *An allele represented by i individuals in a sample of size m is the oldest allele in the sample with probability i/m.*

Proof. The probability in question is just the probability that an individual chosen at random from the population is of the given allelic type, conditional on the individual chosen having an allelic type which is represented in the sample. The probability that a randomly chosen individual is of the given allelic type is

$$\frac{i(\nu+M)}{M(\nu+m)}$$

and the probability that it is of an allelic type represented in the sample is

$$\frac{m(\nu+M)}{M(\nu+m)}$$

Hence the conditional probability sought is i/m.

Exercises 7.2

1. If an allele is represented in the population by j individuals show that the probability the age of the allele is less than x, $P_j(x)$, satisfies

$$\frac{dP_j(x)}{dx}=q(j,j-1)[P_{j-1}(x)-P_j(x)]-q(j,j+1)[P_j(x)-P_{j+1}(x)]$$

where $q(j,j-1)$ and $q(j,j+1)$ are given by equations (7.8). Obtain a recursion for the probability that there have been exactly a births since the allele first appeared.

2. Suppose that $(f_1, f_2, \ldots, f_K)$ gives the frequencies of the alleles present in the population in order of the ages of the alleles, so that f_1 and f_K are the frequencies of the newest and oldest alleles respectively. Thus $(f_1, f_2, \ldots, f_K)$ contains rather more information than **M**. Show that the age of the oldest allele is independent of $(f_1, f_2, \ldots, f_K)$.
3. Show that expression (7.9) is increasing or decreasing in i according to whether ν is less than or greater than unity. Show that in equilibrium the expected number of individuals of the oldest allelic type is $(\nu+M)/(\nu+1)$.
4. Establish Corollary 7.8 for a sample from the model of Exercise 7.1.10.
5. If alleles $1, 2, \ldots, k$ are represented in a sample by $n_1, n_2, \ldots, n_k$ individuals respectively, show that the probability allele r is older than allele $r+1$, for $r = 1, 2, \ldots k-1$, is

$$\prod_{r=1}^{k} \frac{n_r}{\sum_{i=r}^{k} n_i}$$

6. A sample of size m is to be taken from a population in equilibrium. Deduce from Corollary 7.8 that the probability the oldest allele in the sample will be represented by i individuals in the sample is given by expression (7.9) with M replaced by m.
7. We have seen that, for the reversible model considered, if we observe the state of the process at a fixed point in time the age of an allele present in the population has the same distribution as the time to extinction of that same allele. Consider now the labelled process $(\mathbf{M}, \mathbf{l})$ arising from the non-reversible genetic model of Exercise 7.1.14; $2M$ labels are required to ensure an interval between successive uses of the same label for different alleles, and it is simplest to suppose that at each point in (discrete) time the labels $l(j, 1), l(j, 2), \ldots, l(j, M_j)$ are the labels of those alleles represented by j individuals in the population arranged in a random order. Use Exercise 1.4.3 to show that if we do *not* observe the state of the process then at a fixed point in time the age of an allele given to be present in the population has the same distribution as the time to extinction of that same allele. Observe that this result follows from the stationarity of the model; the stronger property of reversibility is only required if we are given information about the state of the process at the fixed point in time.

7.3 FIXATION TIMES

If the mutation rate u is low it is quite likely that every individual in the population will be of the same allelic type. Recurrent mutation ensures, however, that no allele can become permanently fixed in the population. Call an allele *quasi-fixed* if it is the only allele present in the population. We shall begin this section by calculating the probability that an allele becomes quasi-fixed and also the mean time between quasi-fixations of different alleles.

Theorem 7.9. *The probability that a new allele will become quasi-fixed is Q where*

$$Q^{-1}=\sum_{i=0}^{M-1}\binom{M-1}{i}^{-1}\binom{\nu+M-1}{i}$$

Proof. When a new allele appears it will be represented by one individual in a population of M. The future frequency of the allele in the population will follow a random walk with transition intensities (7.8); we are interested in the probability that this random walk reaches M before it reaches zero. This can be determined by the standard means for obtaining absorption probabilities or by using the electrical analogue described in Section 5.2. In the electrical analogue the resistance between nodes i and $i+1$ is proportional to

$$\binom{M-1}{i}^{-1}\binom{\nu+M-1}{i}$$

and so if nodes 0 and M are held at potentials of 0 and 1 respectively the potential of node 1 will be Q, establishing the theorem.

When $\nu=1$ the expression for Q^{-1} becomes $M(1+\frac{1}{2}+\cdots+1/M)\simeq M\log M$. As ν approaches zero, Q^{-1} approaches M.

Corollary 7.10. *The intervals between first quasi-fixations of different alleles are independent and identically distributed with mean $(QM\mu u)^{-1}$.*

Proof. The same allele may become quasi-fixed more than once during its existence, but if we consider only those points in time at which an allele becomes quasi-fixed for the first time then it is clear that the intervening intervals are independent and identically distributed. Let the mean interval length be x. Now the mean period between successive mutations is $(M\mu u)^{-1}$, and so the long-run proportion of alleles which become quasi-fixed must be $(M\mu ux)^{-1}$. But this long-run proportion is Q by the previous theorem, and hence $x=(QM\mu u)^{-1}$.

When a mutation occurs the new allele will generally bear a close resemblance to the allelic type of the mutant's parent. This is because an allele is made up of a large number of smaller units, called nucleotides, and when a mutation occurs it is unlikely to affect more than one of these units.

Suppose then that an allele is made up of an infinite number of nucleotides and that a mutation affects just one of these nucleotides, with no nucleotide ever affected more than once by a mutation. Thus a mutation gives rise to a new allele, which will eventually disappear from the population, and a new nucleotide, which may or may not disappear from the population. Indeed, since the population will eventually consist entirely of the descendents of one of the M individuals alive at a given time the probability that a new nucleotide will eventually be fixed in the population is M^{-1}. We shall devote the rest of this section to the elucidation of this process of gene substitution.

Call the birth of a mutant individual a *determining mutation* if the entire population will eventually be descended from that individual. Thus a determining mutation gives rise to a nucleotide which will eventually become fixed in the population.

Theorem 7.11. *Determining mutations form a Poisson process of rate μu.*

Proof. We started this chapter by obtaining the infinite alleles model as a limiting case of the reversible migration processes of the last chapter. It is also possible to view the infinite alleles model as an invasion process of the form discussed in Chapter 5: regard the M individuals as sites and when an individual gives birth to an offspring regard this as one site invading another. Each time an invasion occurs there is a probability u that the invaded site becomes a completely new colour, corresponding to a mutation. Viewed in this light it is natural to choose an individual alive at time 0 and trace the ancestry of this individual. As we move backwards through time the points in time at which his ancestors were born form a Poisson process of rate μ. Further, the points in time at which an ancestor of his was born a mutant form a Poisson process of rate μu. This is true for each of the M individuals alive at time 0. We thus have M (dependent) Poisson processes; one of them is the sequence of determining mutations up until time 0, but we do not know which one since we cannot tell from which of the M individuals alive at time 0 the population will eventually be descended. Nevertheless, it will be one of these processes and since they are all Poisson processes of rate μu the theorem is proved.

Thus nucleotides destined to be fixed in the population arise as a Poisson process, and the number of nucleotides by which an individual alive at time t differs from his ancestor at time 0 is Poisson with mean μut. The points in time at which nucleotides actually become fixed form a much more complicated point process (Exercise 7.3.4).

Exercises 7.3

1. Show that the mean time between first quasi-fixations tends to infinity when the mutation rate u tends to either zero or infinity (with the population size M held fixed).

2. Show that the expected time till a new nucleotide is fixed or lost is

$$\frac{M-1}{\mu}\sum_{i=1}^{M-1}\frac{1}{i}$$

3. Show that the frequency of a nucleotide destined to be fixed performs a random walk with transition intensities

$$q(i, i-1) = \mu\frac{(i-1)(M-i)}{M(M-1)}$$

$$q(i, i+1) = \mu\frac{(i+1)(M-i)}{M(M-1)}$$

Prove that for a new nucleotide destined to be fixed the expected time till fixation is

$$\frac{(M-1)^2}{\mu}$$

and for a new nucleotide destined to be lost the expected time till loss is

$$\frac{M}{\mu}\sum_{i=2}^{M}\frac{1}{i}$$

4. Observe that arbitrarily many nucleotides may become fixed in the population at the same moment. Show that if $X(t)$ is the number of nucleotides which become fixed in the population in the interval $(0, t)$ then

$$E[X(t)] = \mu ut$$

and

$$\text{Var}[X(t)] = \mu ut + 0(1)$$

where $0(1)$ is a term which remains bounded as $t \to \infty$.
5. Suppose that two distinct individuals are chosen at random from the population. Show that the number of nucleotides by which they differ has the same distribution as $X = Y + Z$ where Y is geometric with mean $(M-2)u$, Z is Bernoulli with mean u, and Y and Z are independent.
6. Suppose that at time 0 a population of size M is subdivided into a number of colonies, which henceforth evolve separately from each other with the original values of μ and u. Show that two individuals chosen at random from distinct colonies at time t differ by a number of nucleotides which has the same distribution as $W + X$ where W has a Poisson distribution with mean $2\mu ut$, X is as in the previous exercise, and W and X are independent.

7. The proof of Theorem 7.11 does not depend upon the reversibility of the genetic model considered. Show that in the non-reversible genetic model of Exercise 7.1.14 the intervals between successive determining mutations are independent and have the same distribution as $Y+1$ where Y has a geometric distribution.

CHAPTER 8

Clustering Processes

Processes in which individuals or units form themselves into clusters have applications in a variety of fields. In Section 8.1 we shall commence investigation of these clustering processes with an example arising from the study of social behaviour. Later, in Section 8.4, we shall consider in some detail a more complex example taken from the field of polymer chemistry.

8.1 INTRODUCTION

In Section 6.2 we considered a model for social grouping behaviour based on the reversible migration processes of that chapter. A possible drawback of the model was its assumption that groups form at specific locations; this assumption might be appropriate for troops of monkeys sleeping in trees, but is less likely to apply to clusters of people at a cocktail party. An alternative model for the latter context might be as follows. Suppose there are M individuals formed into distinct groups, with m_i being the number of groups consisting of i individuals. Thus

$$\sum_{i=1}^{M} im_i = M \tag{8.1}$$

and m_1 is the number of isolated individuals, or isolates. Suppose that a given isolate moves to join a given group of size $i(i=1, 2, \ldots)$ at rate α and that a given individual within a group of size $i(i=2, 3, \ldots)$ leaves that group to become an isolate at rate β. Put more precisely this is equivalent to the assumption that $\mathbf{m}=(m_1, m_2, \ldots)$ is a Markov process in which the transition rate from $(m_1, m_2, \ldots, m_i, m_{i+1}, \ldots)$ to $(m_1-1, m_2, \ldots, m_i-1, m_{i+1}+1, \ldots)$ is $\alpha m_1 m_i$ for $i \geq 2$; from $(m_1, m_2, m_3, \ldots)$ to $(m_1-2, m_2+1, m_3, \ldots)$ is $\alpha m_1(m_1-1)$; from $(m_1, m_2, \ldots, m_{i-1}, m_i, \ldots)$ to $(m_1+1, m_2, \ldots, m_{i-1}+1, m_i-1, \ldots)$ is $i\beta m_i$, $i>2$; and from $(m_1, m_2, m_3, \ldots)$ to $(m_1+2, m_2-1, m_3, \ldots)$ is $2\beta m_2$. Note that two isolates may form a group of size 2; the rate at which a group of size 2 is formed in this way while there are m_1 isolates is $\alpha m_1(m_1-1)$, there being $m_1(m_1-1)$ distinct ordered pairs of isolates.

The equilibrium distribution is

$$\pi(\mathbf{m}) = B \prod_{i=1}^{M} \frac{1}{m_i!} \left(\frac{\beta}{\alpha i!}\right)^{m_i} \tag{8.2}$$

for $\mathbf{m}$ satisfying restriction (8.1), where B is a normalizing constant. This can be verified by checking the detailed balance conditions, and is a special case of a more general result to be discussed in the next section. We have considered this special case separately in order to motivate the next two sections. We shall see later (Exercise 8.2.1) that from the equilibrium distribution it is possible to calculate quantities of interest such as the expected number of groups of size i or the expected proportion of individuals in groups of size i.

The process $\mathbf{m}$ has some similarities with the infinite alleles model $\mathbf{M}$ considered in the previous chapter. In each case the ith component gives the number of groups of i individuals. Like the infinite alleles model the simple clustering model described in this section can be considered as a limiting case of a reversible migration process (Exercise 8.1.1).

Exercises 8.1

1. The process described in this section is a limiting form of a reversible migration process. To see this consider a reversible migration process $(n_1, n_2, \ldots, n_J)$ with J sites and $\sum_{j=1}^{J} n_j = M$, and suppose the probability intensity an individual moves from site j to site k is $\phi(n_j)$ where $\phi(1) = \alpha$ and $\phi(n) = \beta n/J$, $n \geq 2$. Now let $\mathbf{m} = (m_1, m_2, \ldots)$ describe the system with m_i being the number of sites inhabited by i individuals. Show that as $J \to \infty$ the transition rates and the equilibrium distribution for the process $\mathbf{m}$ approach those of the model described in this section.
2. If $\mathbf{m}$ is distributed according to (8.2) with α/β an unknown parameter show that the number of groups, $\sum m_i$, is a sufficient statistic for α/β.

8.2 THE BASIC MODEL

In this section we shall discuss a fairly general clustering process. It will come as no surprise that when the process is reversible equilibrium results can be readily obtained.

Suppose there exist a countable number of possible cluster types, labelled $r = 1, 2, \ldots$. Two clusters may join together to form a single cluster or a cluster may break up into two clusters. Let m_r be the number of r-clusters, i.e. clusters of type r, and let $\mathbf{m} = (m_1, m_2, \ldots)$. Define the operators R_u^{rs} and R_{rs}^u by

$$R_u^{rs}\mathbf{m} = (m_1, m_2, \ldots, m_r - 1, \ldots, m_s - 1, \ldots, m_u + 1, \ldots)$$

and

$$R_{rs}^u\mathbf{m} = (m_1, m_2, \ldots, m_r + 1, \ldots, m_s + 1, \ldots, m_u - 1, \ldots)$$

so that they correspond respectively to the union of an r-cluster and an s-cluster to form a u-cluster and the break up of a u-cluster into an r-cluster and an s-cluster. Similarly define

$$R_u^{rr}\mathbf{m} = (m_1, m_2, \ldots, m_r - 2, \ldots, m_u + 1, \ldots)$$

and

$$R_{rr}^{u}\mathbf{m} = (m_1, m_2, \ldots, m_r + 2, \ldots, m_u - 1, \ldots)$$

We shall suppose that the process $\mathbf{m}$ is Markov with transition rates

$$\begin{aligned} q(\mathbf{m}, R_u^{rs}\mathbf{m}) &= \lambda_{rsu} m_r m_s \qquad r \neq s \\ q(\mathbf{m}, R_u^{rr}\mathbf{m}) &= \lambda_{rru} m_r (m_r - 1) \\ q(\mathbf{m}, R_{rs}^{u}\mathbf{m}) &= \mu_{rsu} m_u \end{aligned} \tag{8.3}$$

where $\lambda_{rsu} = \lambda_{sru}$ and $\mu_{rsu} = \mu_{sru}$. The parameter μ_{rsu} can be regarded as the probability intensity a given u-cluster breaks up into an r-cluster and an s-cluster; similarly, λ_{rsu} for $r \leq s$ can be regarded as the probability intensity a given r-cluster attaches itself to a given s-cluster to form a u-cluster. Let $\mathcal{S}$ be the state space of the process $\mathbf{m}$. For the moment we shall assume $\mathcal{S}$ is finite and irreducible. In this case call $\mathbf{m}$ a closed clustering process.

Theorem 8.1. *If there exist positive numbers $c_1, c_2, \ldots$ satisfying*

$$c_r c_s \lambda_{rsu} = c_u \mu_{rsu} \tag{8.4}$$

then in equilibrium the closed clustering process $\mathbf{m}$ is reversible with equilibrium distribution

$$\pi(\mathbf{m}) = B \prod_r \frac{c_r^{m_r}}{m_r!} \tag{8.5}$$

where B is a normalizing constant.

Proof. The detailed balance conditions

$$\pi(\mathbf{m}) q(\mathbf{m}, R_u^{rs}\mathbf{m}) = \pi(R_u^{rs}\mathbf{m}) q(R_u^{rs}\mathbf{m}, \mathbf{m}) \tag{8.6}$$

are readily verified; the theorem follows from these.

Often the clusters are made up of basic units which cannot be created or destroyed. In this case if $k(r)$ is the number of units in an r-cluster then $c_r \theta^{k(r)}$, $r = 1, 2, \ldots$, will also satisfy equation (8.4). The distribution (8.5) is, however, unaltered—a factor θ^M is incorporated into the constant B, where M is the total number of units in the system. In the example described in the last section $k(r) = r$ and so the number of units, or individuals, in a cluster defines it completely. This will often be the case but a process for which it is not true will be discussed in Section 8.4. The normalizing constant B is, as in closed queueing systems, straightforward but tedious to calculate (Exercise

8.2.1). As might be expected we can avoid difficulties with B by making the system open. One way to do this is as follows. Define the operators

$$R_{1}^{\cdot}\mathbf{m} = (m_1 + 1, m_2, m_3, \ldots)$$
$$R^{1}_{\cdot}\mathbf{m} = (m_1 - 1, m_2, m_3, \ldots)$$

and suppose that one-clusters may enter or leave the system with corresponding transition rates

$$\begin{aligned} q(\mathbf{m}, R_{1}^{\cdot}\mathbf{m}) &= \nu \\ q(\mathbf{m}, R^{1}_{\cdot}\mathbf{m}) &= \mu m_1 \end{aligned} \tag{8.7}$$

Assume that every state $(m_1, m_2, \ldots)$ with $\sum m_r$ finite can be reached from every other such state.

Theorem 8.2. *If there exist positive numbers $c_1, c_2, \ldots$ satisfying*

$$\begin{aligned} \nu &= c_1 \mu \\ \lambda_{rsu} c_r c_s &= c_u \mu_{rsu} \end{aligned} \tag{8.8}$$

and

$$\sum_{r=1}^{\infty} c_r < \infty \tag{8.9}$$

then the open clustering process $\mathbf{m}$ *with transition rates* (8.3) *and* (8.7) *has equilibrium distribution*

$$\pi(\mathbf{m}) = \prod_{r=1}^{\infty} e^{-c_r} \frac{c_r^{m_r}}{m_r!} \tag{8.10}$$

In equilibrium $\mathbf{m}$ *is reversible and $m_1, m_2, \ldots$ are independent random variables; m_r has a Poisson distribution with mean c_r.*

Proof. For the open process there is in addition to conditions (8.6) the extra detailed balance condition

$$\pi(\mathbf{m})\nu = \pi(R_{1}^{\cdot}\mathbf{m})\mu(m_1 + 1)$$

and it is readily checked that the form (8.10) satisfies all of these equations. The additional assumption (8.9) is necessary to ensure that the form (8.10) assigns probability one to the countable set $\mathscr{S}$ consisting of states $\mathbf{m}$ with $\sum m_r$ finite.

It is often not apparent from the transition rates (8.3) whether equations (8.4) have a non-zero solution; the following lemma, a partial converse to Theorem 8.1, can sometimes be used to establish this.

Lemma 8.3. *Let* $\mathbf{m}$ *be a closed clustering process with transition rates (8.3), and suppose the parameters* μ_{rsu} *permit a sequence of dissociations by which an r-cluster can break up into one-clusters. If* $\mathbf{m}$ *is reversible then there exists a solution to equations (8.4) with the property that* $c_r > 0$ *if in equilibrium there is a positive probability that an r-cluster is present.*

Proof. Let $\mathbf{m}^1$ be a state in which $m_r > 0$, and let $\mathbf{m}^1, \mathbf{m}^2, \ldots, \mathbf{m}^{k(r)}$ be a sequence of states produced by the dissociation of an r-cluster into one-clusters. Thus $\mathbf{m}^{k(r)}$ is the state obtained from $\mathbf{m}^1$ by removing an r-cluster and replacing it with $k(r)$ one-clusters. The condition on the parameters μ_{rsu} ensures that there exists such a sequence with $q(\mathbf{m}^i, \mathbf{m}^{i+1}) > 0$ for $i = 1, 2, \ldots, k(r)-1$. Define

$$c_r = \frac{m_r m_1!}{(m_1 + k(r))!} \prod_{i=1}^{k(r)-1} \frac{q(\mathbf{m}^{i+1}, \mathbf{m}^i)}{q(\mathbf{m}^i, \mathbf{m}^{i+1})}$$

Observe that the first factor has been chosen so that terms involving $m_1, m_2, \ldots$ cancel, leaving c_r as the product of ratios λ_{rsu}/μ_{rsu}. Further, since $\mathbf{m}$ is reversible Kolmogorov's criteria establishes that c_r does not depend on the sequence of states $\mathbf{m}^1, \mathbf{m}^2, \ldots, \mathbf{m}^{k(r)}$ used to define it. Note that c_r will be positive unless the numerator is zero, in which case an r-cluster could not exist in equilibrium. It now remains to check that c_r, $r = 1, 2, \ldots$, satisfy equations (8.4). This can be most readily verified by defining c_u using the sequence $\mathbf{m}^1, \mathbf{m}^2, \ldots$ in which after the u-cluster has separated into an r-cluster and an s-cluster the r-cluster completely breaks up into one-clusters before the s-cluster begins to dissociate.

Exercises 8.2

1. Consider a closed clustering process in which the total number of units present in the system is fixed at M, with r units present in each r-cluster. Let B_M be the normalizing constant appearing in expression (8.5). Thus

$$B_M^{-1} = \sum_{\mathbf{m}} \prod_r \frac{c_r^{m_r}}{m_r!}$$

where the summation runs over all $\mathbf{m}$ satisfying $\sum r m_r = M$. Show that the expected number of clusters of size r is $c_r B_M / B_{M-r}$ and that the probability a randomly chosen unit lies in a cluster of size r is $r c_r B_M / M B_{M-r}$. Deduce the recursive formula for $B_1, B_2, \ldots,$

$$M B_M^{-1} = \sum_{r=1}^{M} r c_r B_{M-r}^{-1} \tag{8.11}$$

where $B_0 = 1$.

2. For the process considered in the previous exercise show that B_M^{-1} is the

coefficient of θ^M in the power series expansion of

$$G(\theta)=e^{F(\theta)}=\exp\left(\sum_r c_r\theta^r\right)$$

Deduce the recursion (8.11) from the identity

$$\theta G'(\theta)=G(\theta)\sum_r rc_r\theta^r$$

3. There are many ways a closed clustering process can be amended to produce an open clustering process. For example, suppose that r-clusters may enter or leave the system at rates

$$\begin{aligned} q(\mathbf{m}, R_r\mathbf{m}) &= \nu_r \\ q(\mathbf{m}, R^r\mathbf{m}) &= \mu_r m_r \end{aligned} \tag{8.12}$$

Show that if

$$\nu_r = c_r\mu_r$$

and

$$c_r c_s \lambda_{rsu} = c_u \mu_{rsu}$$

then the system with transition rates (8.3) and (8.12) is reversible with equilibrium distribution (8.10), provided condition (8.9) holds and any state with $\sum m_r$ finite can be reached from any other such state.

4. Suppose that an r-cluster contains $k(r)$ units. Show that if c_r, $r=1,2,\ldots,$ is a solution to equations (8.4) then $c_r\theta^{k(r)}$, $r=1,2,\ldots,$ is a solution to equations (8.8) where $\theta^{k(1)}=\nu/\mu c_1$.
5. Even when clusters are made up of units the condition imposed on the dissociation rates in Lemma 8.3 may not be satisfied. For example if an r-cluster contains r units and if $\lambda_{rsu}=\mu_{rsu}$, $\lambda_{112}=\lambda_{224}=\lambda_{246}=\lambda_{336}=1$ with all other $\lambda_{rsu}=0$ then a three-cluster can only be formed by the breakup of a six-cluster. In fact the condition imposed in Lemma 8.3 is unnecessary. Show that if a process $\mathbf{m}$ with transition rates (8.3) is reversible then there exists a non-zero solution $c_1, c_2, \ldots$ to equations (8.4).
6. Suppose that the function $F(\theta)$ defined in Exercise 8.2.2 has radius of convergence θ_0. Deduce that $G(\theta)$ has the same radius of convergence, and that if B_M/B_{M-1} tends to a limit as $M\to\infty$ the limiting value is θ_0. Consider now the effect of letting $M\to\infty$ on the equilibrium distribution of the clustering process. Deduce from Exercise 8.2.1 that if B_M/B_{M-1} converges then the expected number of r-clusters tends to $c_r\theta_0^r$.
7. Generalize the results contained in Exercises 8.2.1, 8.2.2, and 8.2.6 to allow an r-cluster to contain $k(r)$ units.

8.3 EXAMPLES

In general the rates λ_{rsu} and μ_{rsu} will not allow a solution to equations (8.4): in this section we shall give some examples where they do. In most of these examples the number of units in a cluster will define it completely, with $k(r)=r$. Initially we shall consider clustering processes in which only a single unit can join on to or break off from a cluster, and we shall write λ_r and μ_r for $\lambda_{1,r,r+1}$ and $\mu_{1,r-1,r}$ respectively. For such processes equations (8.4) always have a solution: it can be built up from the recursion

$$c_r\mu_r = c_{r-1}c_1\lambda_{r-1} \tag{8.13}$$

with c_1 chosen arbitrarily. The solution to equations (8.8) is given by the same recursion, with c_1 set equal to ν/μ.

Example 1. The simplest case occurs when

$$\lambda_r = \alpha \qquad \mu_r = \beta$$

so that size does not affect the propensity of clusters to associate or dissociate. From the recursion (8.13) we find that a solution to equations (8.4) is

$$c_r = \frac{\beta}{\alpha}$$

Thus for a closed clustering process in which the total number of units is fixed at M the equilibrium distribution is

$$\pi(\mathbf{m}) = B_M\left(\frac{\beta}{\alpha}\right)^{\Sigma_r m_r}\prod_r \frac{1}{m_r!} \tag{8.14}$$

where B_M can be calculated from the recursion (8.11). Observe that if β is much smaller than α then we can expect $\sum m_r$ to be small and hence we can expect the M units to be concentrated in a few large clusters. Suppose now the process is made open using transition rates (8.7). The solution to equations (8.8) is $c_r = (\beta/\alpha)\theta^r$ where

$$\theta = \frac{\alpha\nu}{\beta\mu}$$

(Exercise 8.2.4). Define formally the function

$$F(\theta) = \sum_{r=1}^{\infty} c_r$$

so that for this example

$$F(\theta) = \sum_{r=1}^{\infty} \frac{\beta}{\alpha}\theta^r$$

To satisfy (8.9) we require $F(\theta)<\infty$ and thus $\theta<1$. Provided $\theta<1$,

$$F(\theta)=\frac{\beta}{\alpha}\frac{\theta}{1-\theta}$$

is the expected number of clusters. As θ approaches unity the expected number of clusters approaches infinity.

Example 2. The case

$$\lambda_r=\alpha \qquad \mu_r=\beta r$$

was considered in Section 8.1; the closed process has equilibrium distribution (8.2). Figure 8.1 plots the expected number of units in clusters of size r against r, for certain values of the parameters. The open process has

$$c_r=\frac{\beta}{\alpha}\frac{\theta^r}{r!}$$

and thus

$$F(\theta)=\frac{\beta}{\alpha}(e^\theta-1)$$

The radius of convergence of the function $F(\theta)$ is infinite: no matter how large θ becomes the open process has an equilibrium distribution.

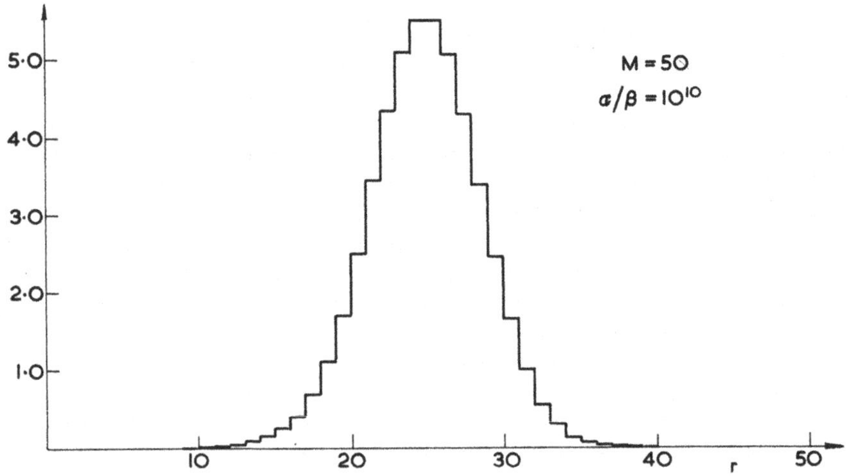

Fig. 8.1 The expected number of units in clusters of various sizes

Example 3. Consider the case

$$\lambda_r=\alpha r \qquad \mu_r=\beta$$

The association rate αr will arise if a given isolate is attracted to each individual of a group at a constant rate α. The closed process has $c_r = (\beta/\alpha)(r-1)!$. The open process does not have an equilibrium distribution for any positive value of θ since $\theta^r(r-1)!$ diverges as r tends to infinity.

Example 4. If

$$\lambda_r = \alpha r \qquad \mu_r = \beta r$$

then the closed process has $c_r = \beta/\alpha r$. For this case the normalizing constant can be written in a closed form and the equilibrium distribution is

$$\pi(\mathbf{m}) = \binom{(\beta/\alpha)+M-1}{M}^{-1} \prod_r \frac{1}{m_r!}\left(\frac{\beta}{\alpha r}\right)^{m_r} \tag{8.15}$$

the same as for the infinite alleles model of Chapter 7. For the open process

$$c_r = \frac{\beta\theta^r}{\alpha r}$$

and so

$$F(\theta) = -\frac{\beta}{\alpha}\log(1-\theta)$$

provided $\theta < 1$.

Example 5. Let

$$\lambda_r = \alpha r \qquad \mu_r = \beta(r+1)$$

The dissociation rate in this and the next example are a little artificial; the purpose of the examples is to introduce a phenomenon which will arise naturally in Section 8.4. For the open process

$$c_r = \frac{2\beta\theta^r}{\alpha r(r+1)}$$

and so an equilibrium distribution exists provided $\theta \leq 1$. As θ approaches unity the expected number of clusters does *not* diverge; in fact the expected number of clusters approaches $F(1) = 2\beta/\alpha$ (Exercise 8.3.1). However, the expected number of units in the system is $\sum rc_r$, and thus as θ approaches unity the expected number of units *does* diverge. When $\theta = 1$ an equilibrium distribution exists, and it has the property that the mean cluster size is infinite (Exercise 8.3.1).

Example 6. Let

$$\lambda_r = \alpha r \qquad \mu_r = \beta(r+2)$$

For the open process

$$c_r = \frac{6\beta\theta^r}{\alpha r(r+1)(r+2)}$$

and so an equilibrium distribution exists provided $\theta \leq 1$. When $\theta = 1$ the expected number of clusters, $\sum c_r$, and the expected number of units, $\sum rc_r$, are both finite. However, $\sum r^2 c_r$ is infinite, a fact which can be interpreted in various ways (Exercise 8.3.2). It shows that the size of a randomly chosen cluster has infinite variance. Suppose there exists a link between two units if they are in the same cluster; thus an r-cluster contains $\frac{1}{2}r(r-1)$ links. When $\theta = 1$ the expected number of links is infinite.

The next examples allow clusters of any size to associate. Write λ_{rs} and μ_{rs} for $\lambda_{r,s,r+s}$, and $\mu_{r,s,r+s}$ respectively. If equations (8.4) have a solution then it can be built up from the recursion

$$c_r \mu_{1,r-1} = c_{r-1} c_1 \lambda_{1,r-1}$$

with c_1 chosen arbitrarily provided $\lambda_{1r} > 0$ for $r \geq 1$. The solution to equations (8.8) is given by the same recursion with c_1 set equal to ν/μ.

Example 7. The obvious generalization of Example 1 is given by the choice of parameters

$$\lambda_{rs} = \alpha \qquad \mu_{rs} = \beta$$

Once again a solution to equations (8.4) is $c_r = \beta/\alpha$, and the equilibrium distribution for the closed system is given by expression (8.14). The open system also has the same equilibrium distribution as for Example 1.

Example 8. One possible generalization of Example 2 which allows a solution to equations (8.4) is obtained by letting

$$\lambda_{rs} = \alpha \qquad \mu_{rs} = \beta\binom{r+s}{r}$$

This dissociation rate arises if the rate at which an $(r+s)$-cluster breaks up to leave a given r units in one cluster and the other s units in another cluster is β, since with the units identified in this way there are $\binom{r+s}{r}$ distinct ways in which an $(r+s)$-cluster can break up to form an r-cluster and an s-cluster. The closed and open equilibrium distributions are the same as for Example 2.

Following the definition of the transition rates (8.3) the parameter λ_{rsu}, $r \leq s$, was interpreted as the probability intensity a given r-cluster attaches itself to a given s-cluster. This interpretation occasionally requires a minor modification when $r = s$. For $r \leq s$ the number of distinct ordered pairs of clusters in which the first cluster is an r-cluster and the second cluster is an s-cluster is $n_r n_s$ (for $r<s$) or $n_r(n_r-1)$ (for $r=s$). However, the number of distinct *unordered* pairs is $n_r n_s$ (for $r<s$) or $\frac{1}{2}n_r(n_r-1)$ (for $r=s$). In the model considered in Section 8.1 (and again as Example 2) it was reasonable to count ordered pairs of clusters with $r \leq s$, and it was therefore reasonable to set $\lambda_{112}=\alpha$. When it is more suitable to count unordered pairs a factor of $\frac{1}{2}$ must be incorporated in the definition of λ_{rru}. To illustrate this consider the following example.

Example 9. Suppose that an r-cluster represents a molecule made up of a string of r atoms arranged along a line with a bond between adjacent atoms. Imagine that the molecules are milling around in a confined space. Molecules combine when they come into close proximity and a molecule breaks into two when a bond breaks. Ignoring spatial considerations it may be reasonable to assume that any given unordered pair of molecules comes into close proximity at rate α and hence to set

$$\lambda_{rs}=\begin{cases}\alpha & r\neq s\\ \frac{1}{2}\alpha & r=s\end{cases}$$

If we suppose that each bond breaks at rate $\frac{1}{2}\beta$ then

$$\mu_{rs}=\begin{cases}\beta & r\neq s\\ \frac{1}{2}\beta & r=s\end{cases}$$

since if $r\neq s$ there are two bonds in an $(r+s)$-cluster whose severance would produce an r-cluster and an s-cluster. The solution to equations (8.4) is $c_r=\beta/\alpha$, and the equilibrium distribution in the open and closed cases are as for Examples 7 and 1.

Occasionally clusters may be constructed from more than one type of basic unit. As an example consider the following elaboration of Example 9.

Example 10. Suppose there are two types of atom, A-atoms and B-atoms. Molecules consist of a chain of atoms with a bond between adjacent atoms. A-atoms can be bonded to up to two other atoms, but B-atoms can only occur at the ends of a chain. Call a chain consisting of just r A-atoms an $r0$-cluster, r A-atoms and one B-atom an $r1$-cluster, and r A-atoms and two B-atoms an $r2$-cluster. Suppose the probability intensity a given (unordered) pair of clusters combine is α if neither of them contain any B-atoms, $\frac{1}{2}\alpha$ if between them they contain one B-atom, and $\frac{1}{4}\alpha$ if they both contain one B-atom. Suppose the rate at which a given bond breaks is $\frac{1}{2}\beta_0$ if it links two A-atoms, $\frac{1}{2}\beta_1$ if it links an A-atom and a B-atom, and $\frac{1}{2}\beta_2$ if it links two B-atoms. It is readily checked that the resulting association and dissociation rates allow the solution

$$c_{r0}=\frac{\beta_0}{\alpha}\qquad c_{r1}=\frac{\beta_0}{\alpha}\qquad c_{r2}=\frac{\beta_0}{4\alpha}\qquad r\geq 1$$

$$c_{01}=\frac{\beta_1}{\alpha}\qquad c_{02}=\frac{\beta_1^2}{4\alpha\beta_2}$$

to equations (8.4).

Exercises 8.3

1. Check that for Example 5,

$$F(\theta)=\frac{2\beta}{\alpha}\left[1+\frac{1-\theta}{\theta}\log(1-\theta)\right]$$

and that $F(1)=2\beta/\alpha$. Show that conditional on the event $\sum m_r=N$ the expected size of a randomly chosen cluster is $\theta F'(\theta)/F(\theta)$, for all $N\geq 1$. Show that when $\theta=1$ the expected size of a cluster chosen at random is infinite. (If $\sum m_r=0$ regard the size of the chosen cluster as zero.)

2. Check that for Example 6,

$$F(\theta)=\frac{3\beta}{2\alpha\theta^2}[3\theta^2-2\theta-2(1-\theta)^2\log(1-\theta)]$$

and that the radius of convergence of $F(\theta)$ is unity. When $\theta=1$ show that the expected number of clusters is $3\beta/2\alpha$, and that, conditional on $\sum m_r>0$, the expected size of a randomly chosen cluster is 2. Show that the expected number of links is $\frac{1}{2}\theta^2F''(\theta)$, and observe that this diverges as $\theta\to 1$. Check that when $\theta=1$ the size of a randomly chosen cluster has infinite variance.

3. The phenomenon observed in Examples 5 and 6 can also arise in quite simple queueing systems. Consider a queue at which arrivals occur at rate $\nu(n+1)/(n+2)$ and customers depart at rate $(n+1)/n$, where n is the number in the queue. Show that an equilibrium distribution exists when $\nu=1$ but the mean number in the queue is infinite.

4. Observe that the equilibrium distribution for Example 8 is unaltered if λ_{rr} is set to $\frac{1}{2}\alpha$ and μ_{rr} to $\frac{1}{2}\beta\binom{2r}{r}$. Show that the resulting dissociation rates arise if u-clusters receive shocks at rate $2^{u-1}\beta$ and if after a cluster has received a shock each of its units assigns itself independently and at random to one of two new clusters. Show that if shocks are received by a u-cluster at rate $f(u)$ then equations (8.4) have a solution only when $f(u)$ takes the form $2^{u-1}\beta$.

8.4 POLYMERIZATION PROCESSES

In this section we shall consider a model of the way in which organic molecules form themselves into polymers. We shall assume that a cluster is constructed from single units linked together by bonds in such a way that the graph formed by regarding units as vertices and bonds as edges is a tree. A cluster breaks up into two clusters by the severance of a single bond, and two clusters join together to form one cluster by the establishment of a single additional bond between two units, one from each cluster. Two clusters are to be regarded as of the same type if the graphs associated with them are isomorphic. Thus when we refer to an r-cluster r identifies the graph associated with the cluster.

Typical clusters are illustrated in Fig. 8.2. There are two simplifications implicit in this model of polymer chains which are worth noting. First, the assumption that clusters correspond to trees rules out cycles, which can of course occur in organic molecules. Second, the assumption that the tree associated with a cluster defines it completely ignores the fact that the angles between bonds are sometimes important.

Suppose that any existing bond breaks at rate κ. This implies that the parameter μ_{rsu} is equal to κ times the number of bonds in a u-cluster whose severance would result in an r-cluster and an s-cluster. The association rate is more difficult to define and arises in the following way. Each unit has f sites (or functionalities, to use the polymer science term) at which a bond can be based. A bond appears at unit rate between any two given vacant sites provided they are on units in different clusters. With this in mind set λ_{rsu} (or $2\lambda_{rru}$ in the case $r=s$) equal to

$$\sum_{x,y} h_x h_y$$

where x and y mark units in an r-cluster and an s-cluster respectively, and the summation runs over all units x and y such that the introduction of a bond between x and y would cause the r-cluster and the s-cluster to form a u-cluster. Let

$$h_x = f - j$$

where j is the number of other units to which x is bonded in the r-cluster and f is a fixed positive integer. Note that the case $f=2$ corresponds to Example 9 of the previous section, with $\alpha = \frac{1}{4}$ and $\beta = 2\kappa$.

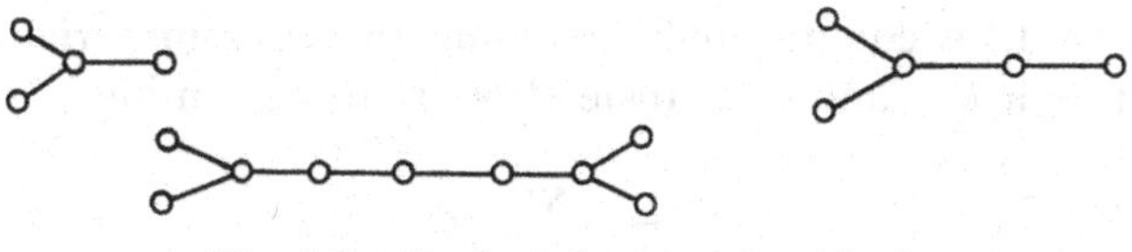

Fig. 8.2 Typical clusters when $f=3$

For given r, s, and u the parameters λ_{rsu} and μ_{rsu} can be explicitly calculated. For example if $f=3$ and r, s, and u are as in Fig. 8.2, then $\lambda_{rsu}=12$ and $\mu_{rsu}=2\kappa$. The precise form of the parameters λ_{rsu} and μ_{rsu} for general r, s, and u involve rather complicated combinational coefficients depending on the shape of r-, s-, and u-clusters. Our first objective will be to show that these parameters allow a non-zero solution to equations (8.4); we shall achieve this by an appeal to Lemma 8.3, after we have shown that the process $\mathbf{m}=(m_1, m_2, \ldots)$, where m_r is the number of r-clusters, is reversible.

Suppose there are in total M units and that these units are distinguishable. Suppose further that each of the f sites on each unit is distinguishable. Redefine the possible cluster types so that the type of a cluster gives not only the graph associated with the cluster but also which unit is at each vertex of the graph, and which sites are used on units for each bond. Use r', s', and u' and primes generally when dealing with clusters described at this level of detail; note that there can exist at most one r'-cluster. Usually it will not be possible for a u'-cluster to be formed from an r'-cluster and an s'-cluster since even if the graphs are compatible the particular units and sites used may not be. In this case $\lambda'_{r's'u'}$ and $\mu'_{r's'u'}$ are both zero. If a u'-cluster can be formed from an r'-cluster and an s'-cluster there will be just one way of doing it; in this case $\lambda'_{r's'u'}=1$ and $\mu'_{r's'u'}=\kappa$. There are no combinatorial coefficients involved in the definition of $\lambda'_{r's'u'}$ and $\mu'_{r's'u'}$, and it is clear that a solution to the equations

$$c'_{r'}c'_{s'}\lambda'_{r's'u'}=c'_{u'}\mu'_{r's'u'}$$

is given by

$$c'_{r'}=\kappa$$

Hence by Theorem 8.1 the process $\mathbf{m}'=(m'_1, m'_2, \ldots)$, where $m'_{r'}$ is the number (zero or one) of r'-clusters, is reversible.

Now $\mathbf{m}$ is a function of $\mathbf{m}'$; m_r is obtained by summing $m'_{r'}$ over all r' whose graph is compatible with r. Hence $\mathbf{m}$ is reversible (Exercise 1.2.9). Further, $\mathbf{m}$ is a Markov process with transition rates of the form (8.3) where λ_{rsu} and μ_{rsu} have been defined earlier. Thus Lemma 8.3 shows that the rates λ_{rsu} and μ_{rsu} allow a non-zero solution $c_1, c_2, \ldots$ to the equations

$$c_r c_s \lambda_{rsu} = c_u \mu_{rsu} \tag{8.16}$$

and we have achieved our first objective.

We shall not obtain a solution $c_1, c_2, \ldots$ to equations (8.16); the expression for c_r would be complicated, involving the combinatorial coefficients appearing through λ_{rsu} and μ_{rsu}. Instead we shall determine $c^*_{r^*}$ where

$$c^*_{r^*} = \sum_{r:k(r)=r^*} c_r$$

We shall see that the mere fact that equations (8.16) have a solution allows us to deduce a recursion for $c^*_{r^*}$ and that from this we can calculate quantities such as the expected number of clusters containing r^* units.

A u-cluster contains $k(u)$ units and hence has $k(u)-1$ bonds. The rate at which a u-cluster breaks up is thus

$$\sum_{r,s} \mu_{rsu} = \kappa(k(u)-1) \tag{8.17}$$

where the summation runs over all unordered pairs (r, s). An r-cluster contains $k(r)$ units, has $k(r)-1$ bonds, and hence has $(f-2)k(r)+2$ vacant sites. The rate at which a given r-cluster and a given s-cluster associate is thus

$$\sum_{u} \lambda_{rsu} = [(f-2)k(r)+2][(f-2)k(s)+2] \qquad r \neq s$$

or (8.18)

$$2\sum_{u} \lambda_{rru} = [(f-2)k(r)+2]^2 \qquad r = s$$

From equation (8.16)

$$c_r\theta^{k(r)}c_s\theta^{k(s)}\lambda_{rsu} = c_u\theta^{k(u)}\mu_{rsu}$$

Summing this over u and unordered pairs (r, s) and substituting from equations (8.17) and (8.18) gives

$$\frac{1}{2}\sum_{r}\sum_{s} c_r\theta^{k(r)}[(f-2)k(r)+2]c_s\theta^{k(s)}[(f-2)k(s)+2]$$

$$= \kappa \sum_{u} c_u\theta^{k(u)}(k(u)-1) \tag{8.19}$$

Notice that the double summation over r and s and the introduction of the factor $\frac{1}{2}$ in equation (8.19) deal adequately with both of the forms appearing in equation (8.18). In terms of the quantities $c^*_{r^*}$ equation (8.19) becomes

$$\frac{1}{2}\sum_{r^*}\sum_{s^*} c^*_{r^*}\theta^{r^*}[(f-2)r^*+2]c^*_{s^*}\theta^{s^*}[(f-2)s^*+2] = \kappa\sum_{u^*} c^*_{u^*}\theta^{u^*}(u^*-1)$$

Equating coefficients of θ^{u^*} produces the recursion

$$\kappa c^*_{u^*}(u^*-1) = \frac{1}{2}\sum_{r^*=1}^{u^*-1} c^*_{r^*}[(f-2)r^*+2]c^*_{u^*-r^*}[(f-2)(u^*-r^*)+2] \tag{8.20}$$

from which $c^*_{r^*}$, $r^* = 1, 2, \ldots$, can be built up. Setting $c^*_1 = \kappa$ the recursion can be shown to produce

$$c^*_{r^*} = \kappa \frac{f^{r^*}[(f-1)r^*]!}{r^*![(f-2)r^*+2]!} \tag{8.21}$$

What is the statistical interpretation of c^*_{r*}? Well we know that $\mathbf{m}$ has the equilibrium distribution $\pi(\mathbf{m})$ given by expression (8.5). Let $\mathbf{m}^* = (m_1^*, m_2^*, \ldots)$ where

$$m^*_{r*} = \sum_{r:k(r)=r^*} m_r$$

the number of clusters of size r^*. The description r^* of a cluster tells us how many units are in the cluster, but no more. If $f>2$ the process $\mathbf{m}^*$ is not Markov; the number of units in a cluster does not give enough information about the various ways the cluster can associate or dissociate. Nevertheless, the equilibrium distribution for $\mathbf{m}^*$ can be calculated from $\pi(\mathbf{m})$ (Exercise 8.4.1) and is given by

$$\pi^*(\mathbf{m}^*) = B \prod_{r^*} \frac{c_{r^*}^{*m^*_{r^*}}}{m^*_{r^*}!} \tag{8.22}$$

for $\mathbf{m}^*$ such that $\sum_i im_i^* = M$. This expression gives the distribution of clusters of various sizes, but does not give more detailed information about the shapes of clusters of a given size. The expected number of clusters of size r^* and the expected number of units in clusters of size r^* can be calculated using Exercise 8.2.1 (see Fig. 8.3).

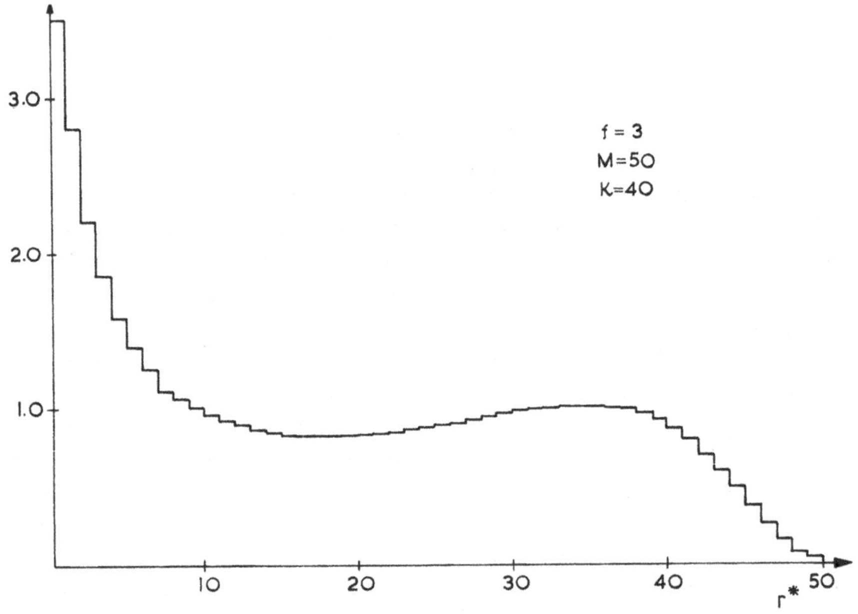

Fig. 8.3 The expected number of units in clusters of various sizes

An important physical phenomenon exhibited by polymers is gelation: when the density is increased above a critical value an observable change takes place in the physical properties of the system. We shall now discuss an elaboration of the above polymerization process which casts some light on this phenomenon. Suppose that a large volume of space consists of J unit sized boxes, and let $\mathbf{m}$ describe the clusters present in a typical box. Suppose that the clusters present in a box associate and dissociate in accordance with rates (8.3) and that in addition each cluster moves from its present box to an adjacent box at rate γ. Suppose, further, that one of the boxes is open to the outside environment with one-clusters emigrating from it and immigrating to it at rates (8.7). The detailed balance conditions allow it to be readily checked that in equilibrium the contents of the J boxes are independent and for each box

$$\pi(\mathbf{m}) = \prod_r e^{-c_r} \frac{c_r^{m_r}}{m_r!}$$

provided conditions (8.8) and (8.9) are satisfied. Thus $m_1, m_2, \ldots$ are independent random variables and m_r has a Poisson distribution with mean c_r. In the polymerization context we can deduce that the number of clusters of size r^* in a typical box has a Poisson distribution with mean

$$c_{r^*}^* = \kappa \frac{(f\theta)^{r^*}[(f-1)r^*]!}{r^*![(f-2)r^*+2]!}$$

where $\theta = \nu/\mu\kappa$, provided $F(\theta) = \sum c_{r^*}^* < \infty$. The radius of convergence of $F(\theta)$ is

$$\theta_0 = \frac{(f-2)^{f-2}}{f(f-1)^{f-1}} \tag{8.23}$$

for $f > 2$ (Exercise 8.4.2). The number of units is independent from box to box with expectation $\sum ic_i^*$. Define the density ρ to be the total number of units divided by the total volume J. For finite J the density is a random variable but from now on we shall consider the limiting case of J infinitely large when the density becomes a constant, $\rho = \sum ic_i^*$. As θ increases towards θ_0 the density increases monotonically to the limiting value

$$\rho_0 = \frac{\kappa(f-1)}{f^2(f-2)}$$

and the sum $\sum i^2c_i^*$ increases monotonically to infinity (Exercise 8.4.6). Consider now how the equilibrium distribution alters as the density is varied. As the density ρ approaches ρ_0 the parameter θ approaches θ_0 and so the expression $\sum i^2c_i^*$ approaches infinity. The divergence of $\sum i^2c_i^*$ corresponds to the physical phenomenon of gelation, whereby as the density passes through a critical value the system moves from the sol to the gel state.

The phenomenon can only occur where $f>2$; when $f=2$ the expressions $\sum ic_i^*$ and $\sum i^2c_i^*$ both increase to infinity as θ approaches θ_0.

Exercises 8.4

1. By summing $\pi(\mathbf{m})$, given by expression (8.5), over all states $\mathbf{m}$ consistent with the reduced description $\mathbf{m}^*$ obtain the distribution (8.22).
2. Use Stirling's formula

$$n! \sim \sqrt{2\pi}n^{n+1/2}e^{-n}$$

(where $\sim$ indicates that the ratio of the two sides tends to unity as $n \to \infty$) to show that the radius of convergence of $F(\theta)=\sum c_r^*$ is given by expression (8.23) for $f>2$. Show that when $\theta=\theta_0$,

$$c_i^* \sim \frac{\kappa(f-1)^{1/2}}{\sqrt{2\pi}(f-2)^{5/2}}\frac{1}{i^{5/2}}$$

3. Show that in the limit of large J the cluster a randomly chosen unit belongs to has mean size $\sum i^2c_i^*/\sum ic_i^*$. It is possible to relate this quantity to the physically observable characteristic, viscosity. The model then predicts that as the density approaches ρ_0 the viscosity becomes infinite.
4. Show that in the limit of large J the probability a randomly chosen site is occupied is

$$\frac{2}{f}\left(1-\frac{F(\theta)}{\theta F'(\theta)}\right)$$

5. To obtain $F(\theta)$ and its derivatives in a closed form it is helpful to change variable from θ to α, where α is the smallest positive root of the equation

$$f\theta = \alpha(1-\alpha)^{f-2}$$

Using this transformation to define α as a function of θ show that

$$\rho = \theta F'(\theta) = \frac{\kappa\alpha}{(1-\alpha)^2f^2}$$

Deduce that

$$F(\theta) = \frac{\kappa\alpha(2-\alpha f)}{2(1-\alpha)^2f}$$

6. Show that the probability a randomly chosen site is occupied, calculated in Exercise 8.4.4, is in fact equal to the variable α of Exercise 8.4.5. Show that as θ approaches θ_0 the probability α approaches $(f-1)^{-1}$. Deduce that as θ increases towards θ_0 the sum $\sum ic_i^*$ increases monotonically to the limiting value ρ_0 and the sum $\sum i^2c_i^*$ increases monotonically to infinity.

7. An alternative approach to the modelling of polymerization yields some results more readily. Suppose that a large vessel contains $2M$ units each with f sites for bonds. From the $2Mf$ sites available select at random $2Mf\alpha$ sites and form these randomly into $Mf\alpha$ pairs of sites. Now introduce a bond between the two sites in each pair. Observe that a proportion α of the sites will be occupied. Suppose now that M is arbitrarily large. Check that the probability a given unit is bonded to itself is zero and that the f sites on a given unit are independently occupied, each with probability α. Check further that the probability r units are linked together by r or more bonds, given they are linked by $r-1$ bonds, is zero. Now choose a unit at random and call the units to which it is directly linked its descendants. Observe that the number of descendants of the initial unit is binomial with parameters f and α, and that each descendant has a further set of descendants, whose number is binomial with parameters $f-1$ and α. Deduce that the expected number of units in the cluster of a randomly chosen unit is

$$\frac{1+\alpha}{1-(f-1)\alpha}$$

Observe that as α increases to $(f-1)^{-1}$ this diverges.

The above model is consistent with that developed in the preceding section. Note that its starting point is α, the proportion of occupied sites, while α is a derived quantity for the model of the preceding section.

8. Suppose that a closed vessel is divided up into J compartments, that in each compartment the clusters present associate and dissociate with rates

$$\begin{aligned} q(\mathbf{m}, R_u^{rs}\mathbf{m}) &= J\lambda_{rsu}m_r m_s \qquad r \neq s \\ &= J\lambda_{rru}m_r(m_r-1) \\ &= \mu_{rsu}m_u \end{aligned}$$

and that each cluster moves from its present box to an adjacent box at rate $\gamma(J)$. Let $\mathbf{m}^+ = (m_1^+, m_2^+, \ldots)$, where m_r^+ is the total number of r-clusters in the vessel. Note that $\mathbf{m}^+$ is not a Markov process. Show that if $c_1, c_2, \ldots$ are positive numbers satisfying equations (8.4) then $\mathbf{m}^+$ has equilibrium distribution given by expression (8.5). Observe that for large J we obtain a model in which clusters perform random walks, possibly merging when they collide.

9. We have used the equilibrium distribution for an open clustering process to investigate the behaviour of a polymerization process as the density approaches the critical value. The open equilibrium distribution is not available when the density exceeds the critical value, but we can use the closed equilibrium distribution. Consider the equilibrium distribution when M units are confined within a fixed finite volume. Exercise 8.2.6

suggests that as $M \to \infty$ the expected number of clusters of size r^* tends to the value $c^*_{r^*}$ obtained with $\theta = \theta_0$. In the limit $(m^*_1, m^*_2, \ldots, m^*_i)$ are independent Poisson random variables for all finite i. Physically this can be interpreted in the following way. As more units are added to the system they increase the gel component but leave unaffected the sol component, which does not depend on density provided this exceeds the critical value. Above the critical point the predictions obtained from the clustering process model and the branching process model of Exercise 8.4.7 differ. When $\alpha > (f-1)^{-1}$ the branching process model relates the expected number of r-clusters to the probability that a unit finds itself in a cluster of size r conditional on the cluster it belongs to being finite, and thus predicts that the sol component depends on α even when this exceeds the critical value. Above the critical density both models are suspect, since the assumption that clusters must correspond to trees becomes untenable.

8.5 GENERALIZATIONS

In this section we shall briefly indicate the results that can be obtained for processes with more general transition rates than those allowed in Section 8.2.

Define the operator $R^u_v\mathbf{m}$ to correspond to the transformation of a cluster of type u into a cluster of type v, so that

$$R^u_v\mathbf{m} = (m_1, m_2, \ldots, m_u - 1, \ldots, m_v + 1, \ldots)$$

Occasionally it may be useful to allow clusters to spontaneously transmute from one form to another. Suppose

$$q(\mathbf{m}, R^u_v\mathbf{m}) = \gamma_{uv} m_u \tag{8.24}$$

It is clear that with an equilibrium distribution of the form (8.5) the detailed balance conditions are satisfied provided $c_1, c_2, \ldots$ satisfy

$$c_u \gamma_{uv} = c_v \gamma_{vu} \tag{8.25}$$

in addition to equations (8.4). To illustrate this consider the following elaboration of Example 7 from Section 8.3, in which $\lambda_{rs} = \alpha$, $\mu_{rs} = \beta$. Suppose that $\gamma_{r,r+1} = \gamma r$ and $\gamma_{r,r-1} = \delta(r-1)$, with $\gamma < \delta$. The first of these rates arises naturally if we suppose that a unit gives birth at rate γ. The second rate is a little contrived, since the obvious choice would be δr. An advantage of $\delta(r-1)$ is that it prevents one-clusters from disappearing, and so allows a non-trivial equilibrium distribution in the absence of immigration. The solution to equations (8.4) and (8.25) is

$$c_r = \frac{\beta}{\alpha}\left(\frac{\gamma}{\delta}\right)^r$$

Any state $\mathbf{m}$ with $\sum m_j < \infty$ is accessible from any other such state; hence in equilibrium $m_1, m_2, \ldots$ are independent, and m_r has a Poisson distribution with mean c_r.

Another extension which springs naturally to mind in view of the migration processes considered in Chapter 2 is the substitution of $\phi_r(m_r)$ for m_r in the transition rates. Consider then the rates

$$\begin{aligned} q(\mathbf{m}, R_u^{rs}\mathbf{m}) &= \lambda_{rsu}\phi_r(m_r)\phi_s(m_s) \qquad r \neq s \\ q(\mathbf{m}, R_u^{rr}\mathbf{m}) &= \lambda_{rru}\phi_r(m_r)\phi_r(m_r - 1) \\ q(\mathbf{m}, R_{rs}^{u}\mathbf{m}) &= \mu_{rsu}\phi_u(m_u) \\ q(\mathbf{m}, R_v^{u}\mathbf{m}) &= \gamma_{uv}\phi_u(m_u) \end{aligned} \tag{8.26}$$

and the possible equilibrium distribution

$$\pi(\mathbf{m}) = B \prod_r \frac{c_r^{m_r}}{\prod_{l=1}^{m_r} \phi_r(l)} \tag{8.27}$$

where it is assumed that B can be chosen to make the distribution sum to unity. It is readily checked that if $c_1, c_2, \ldots$ are a non-zero solution of equations (8.4) and (8.25) then $\pi(\mathbf{m})$ satisfies the detailed balance conditions. Indeed $\pi(\mathbf{m})$ will be the equilibrium distribution under even weaker conditions. If $c_1, c_2, \ldots$ are a non-zero solution of

$$\begin{aligned} c_r c_s \sum_u \lambda_{rsu} &= \sum_u c_u \mu_{rsu} \\ \sum_{r,s} c_r c_s \lambda_{rsu} + \sum_v c_v \gamma_{vu} &= c_u \left(\sum_{r,s} \mu_{rsu} + \sum_v \gamma_{uv} \right) \end{aligned} \tag{8.28}$$

then $\pi(\mathbf{m})$ satisfies the equilibrium equations, even though in this case the process may not be reversible (Exercise 8.5.2).

To illustrate these results consider the following simple example. Suppose that a one-cluster and a two-cluster are single units of different types, and a three-cluster contains one unit of each type. Write λ and μ for λ_{123} and μ_{123} respectively. Then equations (8.4) have the solution $c_1 = c_2 = 1$, $c_3 = \lambda/\mu$, and so the equilibrium distribution is of the form (8.27) whatever the form of the functions ϕ_r, $r = 1, 2, 3$. A special case is

$$\begin{aligned} \phi_1(n) &= \phi_3(n) = n \\ \phi_2(0) &= 0 \\ \phi_2(n) &= 1 \qquad n > 0 \end{aligned} \tag{8.29}$$

This might correspond to a reversible chemical reaction in which the rate of association between molecules of types 1 and 2 is independent of the number of molecules of type 2, provided the number is positive. Alternatively, the model might be appropriate for a social occasion, with three-clusters regarded as dancing couples and the functions ϕ_1 and ϕ_2 reflecting cultural conventions.

In the above example a cluster containing two units survives for an exponentially distributed period before breaking up. Suppose now that such a cluster passes through two stages: call it a three-cluster during the first stage and a four-cluster during the second stage. Consider then a process in which $\lambda_{123}=\lambda$, $\gamma_{34}=2\mu$, $\mu_{124}=2\mu$. Equations (8.28) have the solution $c_1=c_2=1$, $c_3=c_4=\lambda/2\mu$, and so the equilibrium distribution again takes the form (8.27). If $\phi_3(n)=\phi_4(n)=n$ then it is readily shown that in equilibrium (n_1, n_2, n_3+n_4) has the same distribution as had (n_1, n_2, n_3) in the previous process. Indeed the only difference between the two processes is that now the overall lifetime of a cluster containing two units is not exponentially distributed. The method of stages can be used, as in Chapter 3, to obtain results when the overall lifetime of a cluster containing two units is arbitrarily distributed (Exercise 8.5.4).

Exercises 8.5

1. Extend Example 4 of Section 8.3 to allow clusters of any size to associate; show that if $\lambda_{rs}=\alpha rs$ and $\mu_{rs}=\beta(r+s)$ then expression (8.15) remains the equilibrium distribution of the closed process. Suppose now that individuals give birth at rate λ and die at rate μ, so that $\gamma_{r,r+1}=\lambda r$, $\gamma_{r,r-1}=\mu r$, and single individuals immigrate at rate ν. Show that in equilibrium $m_1, m_2, \ldots$ are independent Poisson random variables provided $\alpha\nu=\beta\lambda$.
2. Check that expression (8.27) satisfies the equilibrium equations provided $c_1, c_2, \ldots$ satisfy equations (8.28). Observe that a form of partial balance obtains.
3. Show that if $\lambda_{124}=\mu_{234}=\lambda_{235}=\mu_{125}=1$ with all other λ_{rsu} and μ_{rsu} zero, then equations (8.28) have a solution. Observe that association with a two-cluster transforms a one-cluster into a three-cluster, and vice versa. Show that the process is dynamically reversible.
4. Consider a clustering process with transition rates (8.26) for which equations (8.28) are satisfied and for which the equilibrium distribution has the form (8.27). Suppose that r'-clusters cannot associate with other clusters, that is $\lambda_{r'su}=0$ for all s, u, and that $\phi_{r'}(m)=m$. Thus r'-clusters survive for an exponentially distributed period with mean μ^{-1}, say. Show that the equilibrium distribution $\pi(\mathbf{m})$ will remain the same if the period for which r'-clusters survive has a distribution with mean μ^{-1} which can be expressed as a mixture of gamma distributions.
5. Check that the final example of this section which used the functions (8.29) is equivalent to the finite population telephone exchange model of

Exercise 1.3.5. Observe that allowing three-clusters to survive for an arbitrarily distributed period is equivalent to allowing arbitrarily distributed call lengths, an extension discussed in Section 4.4.

6. In Chapter 6 we considered migration processes in which the rate of migration into colony j was affected by the number of individuals already there, through a function ψ_j. Extend the transition rates (8.26) in an analogous manner and find the equilibrium distribution when equations (8.4) and (8.25) are satisfied.
7. Produce a clustering process analogous to the basic model of Section 8.2, but in which three or more clusters can associate to form, or can result from the dissociation of, a single cluster. Produce the equations analogous to equations (8.4). Develop a finite population telephone exchange model similar to that mentioned in Exercise 8.5.5 but where the subscribers call each other, so that when a call is successfully connected two subscribers and one line become engaged. Show that the form of the equilibrium distribution is unaltered.
8. Consider the garage described in Section 4.6. Suppose the garage and an infinite-server queue form a closed network of queues; interpret the time a car spends in the infinite-server queue as the time between repairs. Describe the resulting system as a clustering process, allowing possible clusters to be a car, a mechanic, or a car plus a mechanic.

CHAPTER 9

Spatial Processes

In this chapter we shall be interested in processes $\mathbf{n}=(n_1, n_2, \ldots, n_J)$ capable of modelling systems containing a finite number of sites or components. The idea is that n_j describes the attribute (or state) of site j and that changes in this attribute are affected by the attributes of sites adjacent to site j. For example sites might be fruit trees in an orchard, and n_j might take the value of unity or zero depending on the presence or absence of disease. In previous chapters the equilibrium distributions obtained have often implied the independence of $n_1, n_2, \ldots, n_J$. Some of the models considered in this chapter lead to a more complicated equilibrium distribution in which there is a limited dependence between $n_1, n_2, \ldots, n_J$. Before discussing these models we shall, in Section 9.1, make precise this concept of limited dependence. Later, in Section 9.4, we shall use the setting provided by these models to discuss the relationship between partial balance and insensitivity.

9.1 MARKOV FIELDS

Consider a system consisting of J sites, each of which has associated with it an attribute. Let n_j be the attribute of site j, where n_j is chosen from a set $\mathcal{N}_j$, assumed for simplicity to be finite. Thus the state of the system, $\mathbf{n}=(n_1, n_2, \ldots, n_J)$, takes values in the state space $\mathcal{S}=\mathcal{N}_1\times\mathcal{N}_2\times\cdots\times\mathcal{N}_J$. A function

$$\pi: \mathcal{S}\rightarrow(0, 1)$$

is called a random field if

$$\sum_{\mathbf{n}\in\mathcal{S}} \pi(\mathbf{n})=1$$

Thus a random field is just a probability distribution over the state space of the system which assigns a positive probability to every state.

To specify the spatial relationship between the sites we shall use a graph theoretic framework. Suppose the J sites of the system are the vertices of a graph G, and call sites j and k neighbours if they are joined by an edge of the graph. Write ∂j for the set of neighbours of j. The same symbol will be used for a graph and its vertex set, and for a site and the set consisting of just that site; thus $G-j$ will refer to the set of sites other than j. For any $H\subset G$ let $|H|$ be the number of sites in H and let $\mathbf{n}_H$ be the $|H|$-dimensional vector giving the value associated with each site in H. Let T_j^m be the operator which changes the attribute of site j to m. Thus

$$T_j^m\mathbf{n}=(n_1, n_2, \ldots, n_{j-1}, m, n_{j+1}, \ldots, n_J)$$

Given a random field π we can calculate the conditional distribution of n_j given the attributes of some or all of the other sites of the system. For example the conditional probability that site j has attribute n_j given that the other sites have attributes $\mathbf{n}_{G-j}$ is

$$P(n_j \mid \mathbf{n}_{G-j})=\frac{\pi(\mathbf{n})}{\sum_{m\in\mathcal{N}_j}\pi(T_j^m\mathbf{n})} \tag{9.1}$$

Conversely for a finite graph G the random field π can be calculated from the conditional probabilities.

Lemma 9.1. *The conditional probabilities $P(n_j \mid \mathbf{n}_{G-j})$, $j\in G$, $\mathbf{n}\in\mathcal{S}$, determine uniquely the random field $\pi(\mathbf{n})$, $\mathbf{n}\in\mathcal{S}$.*

Proof. Let 0 denote one of the attributes from each of the sets $\mathcal{N}_1, \mathcal{N}_2, \ldots, \mathcal{N}_J$, and let $\mathbf{0}=(0, 0, \ldots, 0)$. Observe that the conditional probabilities $P(n_j \mid \mathbf{n}_{G-j})$ determine the ratios $\pi(\mathbf{n})/\pi(T_j^m\mathbf{n})$ since

$$\frac{P(n_j \mid \mathbf{n}_{G-j})}{P(m \mid \mathbf{n}_{G-j})}=\frac{\pi(\mathbf{n})}{\pi(T_j^m\mathbf{n})}$$

Now

$$\frac{\pi(\mathbf{n})}{\pi(\mathbf{0})}=\frac{\pi(\mathbf{n})}{\pi(T_1^0\mathbf{n})}\frac{\pi(T_1^0\mathbf{n})}{\pi(T_1^0T_2^0\mathbf{n})}\cdots\frac{\pi(T_1^0T_2^0\cdots T_{J-1}^0\mathbf{n})}{\pi(T_1^0T_2^0\cdots T_J^0\mathbf{n})}$$

and hence this ratio is also determined by the conditional probabilities. The probability $\pi(\mathbf{0})$ can be deduced from the normalization condition $\sum \pi(\mathbf{n})=1$, and hence the result is proved.

The conditional probabilities $P(n_j \mid \mathbf{n}_{G-j})$ cannot be chosen arbitrarily. They must satisfy certain consistency conditions: the ratio $\pi(\mathbf{n})/\pi(\mathbf{0})$ must not depend on the particular sequence of states $\mathbf{n}$, $T_1^0\mathbf{n}$, $T_1^0T_2^0\mathbf{n}, \ldots, \mathbf{0}$ used to calculate it.

In general $P(n_j \mid \mathbf{n}_{G-j})$ will depend upon the entire vector $\mathbf{n}_{G-j}$, but occasionally it may depend on only some of the components of this vector. Call a random field a *Markov field* if

$$P(n_j \mid \mathbf{n}_{G-j})=P(n_j \mid \mathbf{n}_{\partial j}) \tag{9.2}$$

Thus with a Markov field the conditional probability distribution for the attribute of site j given the attributes of all other sites in the system depends only upon the attributes of sites which are neighbours of site j. Of course if every pair of sites in G is connected by an edge then equation (9.2) is always satisfied. At the other extreme if G contains no edges then condition (9.2) implies that $n_1, n_2, \ldots, n_J$ are independent.

An example of a Markov field is provided if $n_1, n_2, \ldots, n_J$ form successive observations from a realization of a Markov chain. Then

$$\pi(n_1, n_2, \ldots, n_J) = P(n_1)p(n_1, n_2)p(n_2, n_3)\cdots p(n_{J-1}, n_J)$$

and so

$$\begin{aligned} P(n_j \mid \mathbf{n}_{G-j}) &= \frac{p(n_{j-1}, n_j)p(n_j, n_{j+1})}{\sum_m p(n_{j-1}, m)p(m, n_{j+1})} \\ &= P(n_j \mid n_{\partial j}) \qquad 2 \le j \le J-1 \end{aligned}$$

provided we identify the neighbours of j as $j-1$ and $j+1$. This example explains why we use the term Markov field. For a Markov chain $n_1, n_2, \ldots, n_J$ are usually regarded as observations taken at different points in time; here we prefer to regard $n_1, n_2, \ldots, n_J$ as observations taken at different points in space. From a temporal viewpoint the relation

$$P(n_j \mid n_1, n_2, \ldots, n_{j-1}) = P(n_j \mid n_{j-1}) \tag{9.3}$$

is the most natural definition of a Markov chain; we shall see that this is equivalent to the relation

$$P(n_j \mid n_1, n_2, \ldots, n_{j-1}, n_{j+1}, \ldots, n_J) = P(n_j \mid n_{j-1}, n_{j+1}) \tag{9.4}$$

The main result of this section is the next theorem which establishes the form a random field must take if it is to be Markov. To state the theorem we need a little more notation. Call a subset $C \subseteq G$ a simplex if an edge joins any two distinct sites in C, or if C consists of just one site. Let $\mathscr{C}$ denote the set of simplices of the graph G.

Theorem 9.2. *A random field π is a Markov field if and only if it can be written in the form*

$$\pi(\mathbf{n}) = B \prod_{C \in \mathscr{C}} \phi_C(\mathbf{n}_C) \qquad \mathbf{n} \in \mathscr{S} \tag{9.5}$$

Proof. Suppose that π is a Markov field. Write $\mathbf{n}_H^0$ for the J-dimensional vector whose jth component is n_j if $j \in H$ and is zero otherwise. Define the functions $\phi_C(\mathbf{n}_C)$, $C \in \mathscr{C}$, by the recursion

$$\begin{aligned} \phi_j(n_j) &= \pi(\mathbf{n}_j^0) \\ \phi_C(\mathbf{n}_C) &= \frac{\pi(\mathbf{n}_C^0)}{\prod_{H \subset C} \phi_H(\mathbf{n}_H)} \end{aligned} \tag{9.6}$$

Thus $\pi(\mathbf{n})$ takes the form (9.5) with $B=1$ whenever $\mathbf{n}=\mathbf{n}_C^0$ for some $C\in\mathscr{C}$; to prove it always takes this form we shall work by induction on the number of non-zero components in $\mathbf{n}$. If $\mathbf{n}$ is not equal to $\mathbf{n}_C^0$ for some $C\in\mathscr{C}$ then there must be sites j and k which are not neighbours such that n_j and n_k are non-zero. Since π is a Markov field the ratio $\pi(\mathbf{n})/\pi(T_k^0\mathbf{n})$ does not depend upon n_j. Hence

$$\frac{\pi(\mathbf{n})}{\pi(T_k^0\mathbf{n})}=\frac{\pi(T_j^0\mathbf{n})}{\pi(T_j^0T_k^0\mathbf{n})}$$

Write this in the alternative form

$$\pi(\mathbf{n})=\frac{\pi(T_j^0\mathbf{n})\,\pi(T_k^0\mathbf{n})}{\pi(T_j^0T_k^0\mathbf{n})} \tag{9.7}$$

Now $T_j^0\mathbf{n}$, $T_k^0\mathbf{n}$, and $T_j^0T_k^0\mathbf{n}$ all have more zero components than $\mathbf{n}$. Hence our inductive hypothesis allows us to assume $\pi(T_j^0\mathbf{n})$, $\pi(T_k^0\mathbf{n})$, and $\pi(T_j^0T_k^0\mathbf{n})$ are of the form (9.5) with $B=1$. Substitution into equation (9.7) establishes that $\pi(\mathbf{n})$ is also of this form, and so the induction is complete.

The converse is simple to prove. If $\pi(\mathbf{n})$ is of the form (9.5) then

$$\begin{aligned}P(n_j\mid \mathbf{n}_{G-j})&=\frac{\prod_{C\in\mathscr{C}:j\in C}\phi_C(\mathbf{n}_C)}{\sum_{m\in\mathcal{N}_j}\prod_{C\in\mathscr{C}:j\in C}\phi_C(T_j^m\mathbf{n}_C)}\\&=P(n_j\mid \mathbf{n}_{\partial j})\end{aligned}$$

To illustrate the theorem consider the case where the graph G is a finite region of a rectangular lattice and the field is binary, i.e. the attributes of sites are either zero or unity. A site has at most four neighbours, and the only simplices are single sites and pairs of adjacent sites. Suppose that site 1 is internal to the lattice and that its four neighbours are sites 2, 3, 4, and 5. If π is a Markov field the conditional distribution $P(n_1\mid n_{G-1})$ is determined by the ratio

$$\begin{aligned}&\frac{P(n_1=1\mid n_2,n_3,n_4,n_5)}{P(n_1=0\mid n_2,n_3,n_4,n_5)}\\&\quad=\frac{\phi_1(1)\phi_{\{1,2\}}(1,n_2)\phi_{\{1,3\}}(1,n_3)\phi_{\{1,4\}}(1,n_4)\phi_{\{1,5\}}(1,n_5)}{\phi_1(0)\phi_{\{1,2\}}(0,n_2)\phi_{\{1,3\}}(0,n_3)\phi_{\{1,4\}}(0,n_4)\phi_{\{1,5\}}(0,n_5)}\end{aligned}$$

Let α_1 be the value of this ratio when $n_2=n_3=n_4=n_5=0$, and let

$$\beta_{\{1,k\}}=\frac{\phi_{\{1,k\}}(0,0)\phi_{\{1,k\}}(1,1)}{\phi_{\{1,k\}}(1,0)\phi_{\{1,k\}}(0,1)}$$

Then

$$\frac{P(n_1=1\mid n_2,n_3,n_4,n_5)}{P(n_1=0\mid n_2,n_3,n_4,n_5)}=\alpha_1\prod_{k=2}^{5}\beta_{\{1,k\}}^{n_k}$$

Thus a binary Markov field on a rectangular lattice is determined by a relatively small set of parameters: one for each site and one for each pair of adjacent sites. Often symmetry considerations reduce this even further to just two parameters α and β. For an internal site j the conditional probabilities can then be written

$$P(n_j = 1 \mid \mathbf{n}_{G-j}) = \frac{\alpha\beta^r}{1+\alpha\beta^r} \tag{9.8}$$

where r is the number of sites neighbouring j whose attribute is unity. The simplest way to deal with edge effects is to suppose that each lattice point neighbouring the region G is a site which has a known attribute, either zero or unity; expression (9.8) can then be taken to define the conditional probabilities even when j is a site on the boundary of the region, with r including any neighbouring site outside the region whose attribute is unity. The field π can be written as

$$\pi(\mathbf{n}) = B\alpha^M\beta^R \tag{9.9}$$

where $M = \sum n_j$ and R is the number of pairs of neighbouring sites in which the attributes of both sites of the pair are unity.

Exercises 9.1

1. From the recursion (9.6) deduce Grimmett's formula

$$\phi_C(\mathbf{n}_C) = \exp\left(\sum_{H\subseteq C} (-1)^{|C-H|} \log \pi(\mathbf{n}_H^0)\right)$$

2. If $\pi(\mathbf{n})$ is a Markov field the functions ϕ_C appearing in the form (9.5) can be chosen in various ways. For the Markov chain example the obvious choice is

$$\phi_1(n_1) = P(n_1)$$
$$\phi_j(n_j) = 1 \qquad 2 \le j \le J$$
$$\phi_{\{j,j+1\}}(n_j, n_{j+1}) = p(n_j, n_{j+1}) \qquad 1 \le j \le J-1$$

Check that for $2 \le j \le J-1$ Grimmett's formula gives

$$\phi_j(n_j) = P(0)p(0,0)^{J-3}p(0,n_j)p(n_j,0)$$

$$\phi_{\{j,j+1\}}(n_j, n_{j+1}) = \frac{p(n_j, n_{j+1})}{P(0)p(0,0)^{J-2}p(n_j,0)p(0,n_{j+1})}$$

3. Check that for a sequence of random variables $n_1, n_2, \ldots, n_J$ relations (9.3) and (9.4) are equivalent; assume that n_0 and n_{J+1} take known values.
4. Consider a binary Markov field on a graph G which has no simplices

containing more than two sites. Show that if

$$P(n_j = 1 \mid \mathbf{n}_{G-j}) = p_r$$

for all j and $\mathbf{n}_{G-j}$, where r is the number of sites neighbouring j whose attribute is unity, then

$$p_r = \frac{\alpha\beta^r}{1+\alpha\beta^r}$$

for some α and β.

5. Show that for a Markov field the probability distribution $P(\mathbf{n}_H \mid \mathbf{n}_{G-H})$ depends on $\mathbf{n}_{G-H}$ only through the attributes of sites neighbouring H. Associate a graph with the subset H by deleting from the graph G all the sites in $G-H$ and all the edges emanating from these sites. Show that the probability distribution $P(\mathbf{n}_H \mid \mathbf{n}_{G-H})$ is a Markov field over the graph H. Deduce that if H is a tree in which no site has more than two neighbours then, conditional on $\mathbf{n}_{G-H}$, $\mathbf{n}_H$ is a Markov chain. If G is the union of three disjoint sets H_1, H_2, H_3 and if no edge of G joins a site in H_1 to a site in H_3 show that, conditional on $\mathbf{n}_{H_2}$, the random vectors $\mathbf{n}_{H_1}$ and $\mathbf{n}_{H_3}$ are independent.

9.2 REVERSIBLE SPATIAL PROCESSES

Under what conditions will the equilibrium distribution $\pi(\mathbf{n})$ of a stochastic process $\mathbf{n}(t)$ be a Markov field? If changes of attribute at site j are influenced only by the attributes of sites neighbouring j then we might hope that $\pi(\mathbf{n})$ would be a Markov field. However, the transition rates of the invasion processes considered in Section 5.3 have this local character and yet their equilibrium distributions are not Markov fields (Exercise 9.2.1). That a field $\pi(\mathbf{n})$ is Markov is an attractive assumption to make but it is not justified solely by the local character of the transition rates of $\mathbf{n}(t)$. Further restrictions are necessary.

Call $\mathbf{n}(t)$ a *spatial process* if:

(i) Only one component of $\mathbf{n}$ can change at a time.

(ii) The transition rate $q(\mathbf{n}, T_j^m\mathbf{n})$ does not depend on $\mathbf{n}_{G-j-\partial j}$.

(iii) For any states $\mathbf{n}$, $T_j^m\mathbf{n}$ it is possible to reach $T_j^m\mathbf{n}$ from $\mathbf{n}$ by a sequence of transitions which do not alter $\mathbf{n}_{G-j}$.

Condition (iii) can be viewed as a strengthened version of the usual irreducibility assumption. The invasion processes of Section 5.3 are spatial processes provided ν_j and μ_j are positive for all j.

Theorem 9.3. *The equilibrium distribution of a reversible spatial process is a Markov field.*

Proof. The equilibrium distribution $\pi(\mathbf{n})$ satisfies the detailed balance condition

$$\pi(\mathbf{n})q(\mathbf{n}, T_j^m\mathbf{n}) = \pi(T_j^m\mathbf{n})q(T_j^m\mathbf{n}, \mathbf{n})$$

Hence if $q(\mathbf{n}, T_j^m\mathbf{n}) > 0$ the ratio $\pi(\mathbf{n})/\pi(T_j^m\mathbf{n})$ does not depend upon $\mathbf{n}_{G-j-\partial j}$. Condition (iii) ensures that this is true even if it takes more than one transition to reach the state $T_j^m\mathbf{n}$ from $\mathbf{n}$. But these ratios, for $m \in \mathcal{N}_j$, determine the conditional distribution $P(n_j \mid \mathbf{n}_{G-j})$, which is thus equal to $P(n_j \mid \mathbf{n}_{\partial j})$. Hence $\pi(\mathbf{n})$ is a Markov field.

To illustrate the theorem we shall discuss some examples of spatial processes. Suppose that $\mathcal{N}_j = \mathcal{N} = \{1, 2, \ldots, N\}$ and that

$$q(\mathbf{n}, T_j^m\mathbf{n}) = \lambda(n_j, m)\phi(n_j)^r\psi(m)^{r'} \tag{9.10}$$

where r and r' are the numbers of sites neighbouring j which have attributes n_j and m respectively. We can regard $\lambda(n_j, m)$ as the innate tendency of a site's attribute to change from n_j to m, and $\phi(n_j)$ and $\psi(m)$ as measures of the extent to which this tendency is increased or decreased by the existence of neighbouring sites with attributes n_j or m. For example sites may be individuals and attributes may be views on a subject, to give a setting which may help visualize the process. Kolmogorov's criteria readily show that the process is reversible if and only if the rates $\lambda(n, m)$ define a reversible process on the state space $\mathcal{N}$, which happens if and only if there exists a non-zero solution $\alpha(n)$, $n = 1, 2, \ldots, N$, to the equations

$$\alpha(n)\lambda(n, m) = \alpha(m)\lambda(m, n) \tag{9.11}$$

When this is so the equilibrium distribution for the process $\mathbf{n}(t)$ is

$$\pi(\mathbf{n}) = B\prod_{n=1}^{N} \alpha(n)^{M(n)}\left[\frac{\psi(n)}{\phi(n)}\right]^{R(n)} \tag{9.12}$$

where $M(n)$ is the number of sites with attribute n and $R(n)$ is the number of n-bonds, i.e. edges of the graph G which have sites with attribute n at both ends. When $N = 2$ equations (9.11) must have a solution and the equilibrium distribution can be rewritten in the form (9.9).

A drawback of the above model is that the dependence of the transition rates on r, r' is restricted to the multiplicative form given in expression (9.10). The adjective multiplicative is used since if r is increased by one the rate is multiplied by a factor. This form of dependence is typical of processes which have Markov fields as their equilibrium distributions; for reversible processes it is generally a consequence of the detailed balance condition taken together with the multiplicative form enforced by Theorem 9.2.

It is interesting to note that if the functions ϕ, ψ take values close to unity then the rates (9.10) take an approximately additive form. For example suppose $N = 2$, $\lambda(1, 0) = \lambda$, $\lambda(0, 1) = \mu$, $\phi(0) = \phi(1) = \psi(0) = 1$, $\psi(1) = 1 + \delta$.

Consider this process as a model for the ebb and flow of a recurrent infection over an array of plants, with zero indicating a healthy plant and unity indicating the presence of disease. The rate at which an infected plant recovers is λ, and the rate at which a healthy plant becomes infected is

$$\mu(1+\delta)^{r'}$$

where r' is the number of neighbouring plants which are infected. If δ is small this infection rate is approximately equal to

$$\mu+\mu\delta r' \tag{9.13}$$

This is the form we would expect if plants are infected by germs which come from the general environment with intensity μ and from an adjacent infected plant with intensity $\mu\delta$. The approximation will thus be reasonable if germs from the general environment are a significant source of infection. The equilibrium distribution is

$$\pi(\mathbf{n})=B\left(\frac{\mu}{\lambda}\right)^{M(1)}(1+\delta)^{R(1)} \tag{9.14}$$

Our next example of a spatial process has a rather different setting. Suppose the sites are power sources that are connected to power users in such a way that each user has two possible sources of power. Represent the users served by sources j and k as an edge joining sources j and k. Let $d(j, k)$ be the amount of power required by these users; define $d(j, k)$ to be zero if there is no user served by sources j and k. Let n_j be unity or zero according to whether source j is functioning or broken down. If sources j and k are both broken down then demand $d(j, k)$ is unsatisfied. If one of the sources j or k is functioning it supplies the entire demand $d(j, k)$, while if both sources are functioning they each supply an amount $\frac{1}{2}d(j, k)$. Thus if source j is functioning it carries a load $\frac{1}{2}\sum_k (1+n_k)d(j, k)$. If $n_j=1$ let

$$q(\mathbf{n}, T_j^0\mathbf{n})=\lambda_j\exp\left(\tfrac{1}{2}\gamma\sum_k(1+n_k)d(j, k)\right)$$

This breakdown rate corresponds to the fairly severe assumption that each additional unit of load a source has to carry increases its failure rate by a factor e^{γ}. If $n_j=0$ let

$$q(\mathbf{n}, T_j^1\mathbf{n})=\mu_j$$

so that source j remains broken down for a period exponentially distributed with mean μ_j^{-1}. It is readily checked that the detailed balance conditions hold with

$$\pi(\mathbf{n})=B\left\{\prod_{j=1}^{J}\left(\frac{\mu_j}{\lambda_j}\right)^{n_j}\right\}\exp\left(-\tfrac{1}{2}\gamma\sum_{1\le j<k\le J}(1+n_j)(1+n_k)d(j, k)\right)$$

With respect to the graph G formed by linking sources j and k if $d(j, k)>0$ the process $\mathbf{n}$ is a reversible spatial process and the equilibrium distribution is a Markov field. Exercise 9.2.4 discusses a natural extension of this model in which the resulting Markov field involves factors arising from simplices containing more than two sites.

Exercises 9.2

1. For the invasion process of Section 5.3 let $n_j = 0$ or 1 depending on whether site j is white or black. Show that the equilibrium distribution $\pi(\mathbf{n})$ of an invasion process is not in general a Markov field, even when ν_j and μ_j are positive for all j.
2. Our definition of a random field required that it assign positive probability to every state in $\mathscr{S}$. If we remove this restriction we might hope that a field satisfying (9.2) whenever the conditioning events have positive probabilities could be expressed in the form (9.5) or as a limiting case of this form with some of the functions ϕ approaching zero or infinity. This is not so; there exist counterexamples. Similarly, the strong irreducibility condition (iii) is more than a restriction introduced to simplify the proof of Theorem 9.3. Let the graph G consist of three sites with edges joining sites 1 and 2 and sites 2 and 3, and let $\mathcal{N} = \{0, 1\}$. Consider the process (n_1, n_2, n_3) with transition rates as given in Fig. 9.1. Observe that it is a reversible process satisfying conditions (i) and (ii) but not (iii). Show that its equilibrium distribution is not a Markov field.

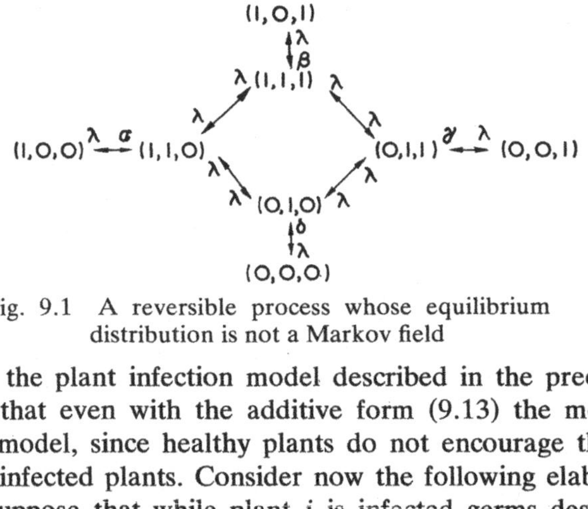

Fig. 9.1 A reversible process whose equilibrium distribution is not a Markov field

3. Consider the plant infection model described in the preceding section. Observe that even with the additive form (9.13) the model is not an invasion model, since healthy plants do not encourage the recovery of adjacent infected plants. Consider now the following elaboration of the model. Suppose that while plant j is infected germs destined to infect plant k are emitted from it at rate $\mu\delta_{jk}$, where $\delta_{jk} = \delta_{kj}$. The value of δ_{jk} might depend on the distance between plants j and k, and could possibly be zero. Show that provided the δ's are small the model approximates a process whose equilibrium distribution is

$$\pi(\mathbf{n}) = B\left(\frac{\mu}{\lambda}\right)^{\Sigma n_j} \prod_{1 \le j < k \le J} (1 + \delta_{jk})^{n_j n_k}$$

4. In the power supply model it was assumed that each user had exactly two sources of power. This assumption can be relaxed. Suppose that for each simplex $C \in \mathscr{C}$ of the graph G a demand $d(C)$ arises from users who can take power from any of the sources in C, and that this demand is shared equally over those sources in C which are functioning. Write down the breakdown rate for source j and deduce that in equilibrium

$$\pi(\mathbf{n}) = B\left\{\prod_{j=1}^{J}\left(\frac{\mu_j}{\lambda_j}\right)^{n_j}\right\}\exp\left[-\gamma \sum_{C\in\mathscr{C}}\left(1+\tfrac{1}{2}+\tfrac{1}{3}+\cdots+\frac{1}{\sum_{k\in C} n_k}\right)d(C)\right]$$

9.3 A GENERAL SPATIAL PROCESS

It is easy to construct reversible spatial processes with a given Markov field as their equilibrium distribution, using the detailed balance condition. In this section we shall describe a fairly general process which gives some insight into the way a Markov field can arise as the equilibrium distribution of a non-reversible process. The closed migration process of Chapter 2 is a special case of the process to be described, and in the next section we shall see how this relationship clarifies the phenomenon of partial balance observed in Chapter 2.

Suppose there are defined positive functions $\Phi(\mathbf{n})$, $\Phi_{G-j}(\mathbf{n}_{G-j})$, $j \in G$. Consider the process $\mathbf{n}(t)$ with transition rates

$$q(\mathbf{n}, T_j^m \mathbf{n}) = \lambda_j(n_j, m)\frac{\Phi(\mathbf{n})}{\Phi_{G-j}(\mathbf{n}_{G-j})} \tag{9.15}$$

We can regard $\lambda_j(n_j, m)$ as the innate tendency of site j to change its attribute from n_j to m, and the other term appearing in the rate (9.15) as a measure of the extent to which this is affected by the attributes of the other sites in the system. We shall assume that for each j the equations

$$\alpha_j(n) \sum_{m\in\mathcal{N}_j} \lambda_j(n, m) = \sum_{m\in\mathcal{N}_j} \alpha_j(m)\lambda_j(m, n) \qquad n \in \mathcal{N}_j \tag{9.16}$$

have a positive solution for $\alpha_j(n)$, $n \in \mathcal{N}_j$, which is unique up to a multiplying factor—this is equivalent to the assumption that the state space $\mathscr{S}$ of the process $\mathbf{n}(t)$ is irreducible.

Theorem 9.4. *The equilibrium distribution for the process $\mathbf{n}(t)$ with transition rates (9.15) is*

$$\pi(\mathbf{n}) = B\frac{\prod_{j=1}^{J} \alpha_j(n_j)}{\Phi(\mathbf{n})} \tag{9.17}$$

where B is a normalizing constant.

Proof. The equilibrium equations are

$$\pi(\mathbf{n})\sum_j\sum_m q(\mathbf{n}, T_j^m\mathbf{n}) = \sum_j\sum_m \pi(T_j^m\mathbf{n})q(T_j^m\mathbf{n}, \mathbf{n}) \qquad \mathbf{n}\in\mathcal{S}$$

By substitution we can verify that $\pi(\mathbf{n})$ satisfies the partial balance equations

$$\pi(\mathbf{n})\sum_m q(\mathbf{n}, T_j^m\mathbf{n}) = \sum_m \pi(T_j^m\mathbf{n})q(T_j^m\mathbf{n}, \mathbf{n}) \qquad \mathbf{n}\in\mathcal{S} \tag{9.18}$$

for each $j\in G$. The equilibrium equations follow from these.

If the function Φ has the appropriate form then $\pi(\mathbf{n})$ will be a Markov field. Whether the process $\mathbf{n}(t)$ is a spatial process will depend on the graph G and on the functions Φ_{G-j}, $j\in G$, as well as on Φ. For example if

$$\Phi(\mathbf{n}) = \prod_{C\in\mathscr{C}} \phi_C(\mathbf{n}_C)$$

$$\Phi_{G-j}(\mathbf{n}_{G-j}) = \prod_{C\in\mathscr{C}:j\in C} \phi_C(\mathbf{n}_C)$$

then

$$q(\mathbf{n}, T_j^m\mathbf{n}) = \lambda_j(n_j, m)\prod_{C\in\mathscr{C}:j\notin C}\phi_C(\mathbf{n}_C) \tag{9.19}$$

so that $\mathbf{n}(t)$ is a spatial process and the equilibrium distribution is the Markov field

$$\pi(\mathbf{n}) = B\frac{\prod_{j=1}^J \alpha_j(n_j)}{\prod_{C\in\mathscr{C}}\phi_C(\mathbf{n}_C)} \tag{9.20}$$

Specializing further suppose $\mathcal{N}_j = \{1, 2, \ldots, N\}$,

$$\phi_C(\mathbf{n}_C) = \begin{cases}\phi(n) & \text{if } |C|=2 \text{ and } \mathbf{n}_C=(n, n)\\ 1 & \text{otherwise}\end{cases}$$

and $\lambda_j(n, m) = \lambda(n, m)$, so that $\alpha_j(n) = \alpha(n)$. Then

$$q(\mathbf{n}, T_j^m\mathbf{n}) = \lambda(n_j, m)\phi(n_j)^r \tag{9.21}$$

where r is the number of sites neighbouring site j which have the same attribute as site j. We could perhaps regard neighbouring sites with the same attribute as forming a bond which decreases, or increases, the rate of change of attribute at those sites. The equilibrium distribution for the process is

$$\pi(\mathbf{n}) = B\prod_{n=1}^N \frac{\alpha(n)^{M(n)}}{\phi(n)^{R(n)}} \tag{9.22}$$

where $M(n)$ is the number of sites with attribute n and $R(n)$ is the number of n-bonds, i.e. edges of the graph G which have sites with attribute n at both ends.

The process with transition rates (9.21) and equilibrium distribution (9.22) is very similar to the process considered in the last section with transition rates (9.10) and equilibrium distribution (9.12). The process considered there imposed a restriction on the parameters $\lambda(n, m)$, but it allowed the transition rates to involve the function $\psi(m)$. The relationship between the two processes is analogous to that between the migration processes of Section 2.3 and the reversible migration processes of Chapter 6.

We shall now show that the closed migration process of Section 2.3 can itself be viewed as a spatial process provided the graph G is taken as the complete graph in which every pair of sites is joined by an edge. Let $\mathcal{N}_j = \{1, 2, \ldots, N\}$ and $\lambda_j(n, m) = \lambda(n, m)$. Let $M(i)$ be the number of sites with attribute i. Thus $\mathbf{M} = (M(1), M(2), \ldots, M(N))$ is a function of $\mathbf{n}$. Let

$$\Phi(\mathbf{n}) = \prod_{i=1}^{N} \frac{1}{M(i)!} \prod_{r=1}^{M(i)} \phi_i(r)$$

and

$$\Phi_{G-j}(\mathbf{n}_{G-j}) = \frac{M(n_j)\Phi(\mathbf{n})}{\phi_{n_j}(M(n_j))}$$

Observe that Φ_{G-j} is indeed a function of $\mathbf{n}_{G-j}$. Substituting these functions into equation (9.15) gives the transition rates of the process $\mathbf{n}$ as

$$q(\mathbf{n}, T_j^m\mathbf{n}) = \lambda(n_j, m)\frac{\phi_{n_j}(M(n_j))}{M(n_j)}$$

The process $\mathbf{M}$ is also Markov and its transition rates take a simpler form. Using the operator T_{ik} introduced in Section 2.3 the process $\mathbf{M}$ has transition rates

$$q(\mathbf{M}, T_{ik}\mathbf{M}) = \lambda(i, k)\phi_i(M(i))$$

It is thus a closed migration process of the form discussed in Section 2.3. The process $\mathbf{M}$ can be viewed as a summary of the information contained in the process $\mathbf{n}$: n_j records the colony which contains individual j and $M(i)$ records the number of individuals in colony i. The transition rates and equilibrium distribution of the process $\mathbf{M}$ take the more natural form. On the other hand, the process $\mathbf{n}$ has the advantage that as a spatial process only one of its components can change at a time; in the next section we shall see that this facilitates a discussion of partial balance. The partial balance equations (2.5) for the closed migration process $\mathbf{M}$ can be deduced from the partial balance equations (9.18) for the spatial process $\mathbf{n}$; the probability flux out of state $\mathbf{M}$ due to an individual moving from colony i is equal to the probability flux into state $\mathbf{M}$ due to an individual moving to colony i, since the probability flux out of state $\mathbf{n}$ due to individual j moving from colony i is equal to the probability flux into state $\mathbf{n}$ due to individual j moving to colony i.

Let us return now to the general process with transition rates (9.15). Consider the period of time for which a site's attribute remains unchanged. We can imagine that after the attribute of site j becomes n_j the attribute has a nominal lifetime exponentially distributed with unit mean which it ages through at rate

$$\lambda_j(n_j)\frac{\Phi(\mathbf{n})}{\Phi_{G-j}(\mathbf{n}_{G-j})}$$

where

$$\lambda_j(n_j)=\sum_m \lambda_j(n_j, m)$$

and that when the attribute's lifetime ends site j takes on attribute m with probability $\lambda_j(n_j, m)/\lambda_j(n_j)$. It is clear that this description of the evolution of the system is consistent with the transition rates (9.15). Now suppose that an attribute's nominal lifetime has some arbitrary distribution with unit mean, where this distribution may vary from attribute to attribute and from site to site. The process $\mathbf{n}(t)$ will no longer be a Markov process, but our experience with migration processes suggests that its equilibrium distribution may still be given by expression (9.17). We shall now give a brief indication of how this can be proved; in the next section we shall see it is a consequence of a more general result. Suppose, to begin with, that all nominal lifetimes are exponentially distributed apart from one attribute at one site. Suppose that at site 1 attribute 1 has as a nominal lifetime the sum of w independent stages, each exponentially distributed with mean w^{-1}. Consider the process $\mathbf{n}'=(n_1', n_2, n_3, \ldots, n_J)$ with $n_1'=n_1$ when $n_1\neq 1$, and with $n_1'=(1, u)$ when $n_1=1$, where the indicator u takes a value between 1 and w depending on which stage of the attribute's lifetime is in progress. Although the process $\mathbf{n}(t)$ is not Markov its value can be deduced from the process $\mathbf{n}'(t)$ which *is* Markov. The transition rates of the process $\mathbf{n}'(t)$ are

$$q(\mathbf{n}', T_j^m\mathbf{n}')=\lambda_j(n_j, m)\frac{\Phi(\mathbf{n})}{\Phi_{G-j}(\mathbf{n}_{G-j})} \tag{9.23}$$

unless $j=1$, in which case the transition rate must be defined more carefully. If $n_1\neq 1$ then the rate at which n_1 changes to $(1, 1)$ is

$$\lambda_1(n_1, 1)\frac{\Phi(\mathbf{n})}{\Phi_{G-j}(\mathbf{n}_{G-j})}$$

If $n_1'=(1, u)$ for $1\leq u\leq w-1$ then the rate at which n_1' changes to $(1, u+1)$ is

$$\lambda_1(1)w\frac{\Phi(\mathbf{n})}{\Phi_{G-j}(\mathbf{n}_{G-j})}$$

If $n_1' = (1, w)$ then the rate at which n_1' changes to m is

$$\lambda_1(1, m)w\frac{\Phi(\mathbf{n})}{\Phi_{G-j}(\mathbf{n}_{G-j})}$$

All other transitions involving site 1 have their rates given by expression (9.23). Thus Theorem 9.4 applies to the process $\mathbf{n}'$, and from this it can be deduced that the equilibrium distribution is

$$\pi'(\mathbf{n}') = B\frac{\alpha_1'(n_1')\prod_{j=2}^{J}\alpha_j(n_j)}{\Phi(\mathbf{n})} \tag{9.24}$$

where

$$\alpha_1'(n_1') = \alpha_1(n_1) \qquad n_1 \neq 1$$

$$\alpha_1'(1, u) = \frac{1}{w}\alpha_1(1)$$

All that needs to be checked is that if α_1 is a solution of equations (9.16) then α_1' is the appropriate solution for the process $\mathbf{n}'$. But now the equilibrium distribution for $\mathbf{n}$ can be obtained by a simple summation of the distribution (9.24). This shows that

$$\pi(\mathbf{n}) = B\frac{\prod_{j=1}^{J}\alpha_j(n_j)}{\Phi(\mathbf{n})}$$

and so we have established the desired result in the case where one attribute has a nominal lifetime with a gamma distribution. At the cost of some additional notation the result can be established when nominal lifetimes of any number of attributes are distributed as mixtures of gamma distributions. As in Section 3.3 this strongly suggests the result for arbitrary distributions, but again the techniques needed for this step are beyond the scope of this work.

To illustrate the result, consider the process with transition rates (9.21) where $n_j = 0, 1$, $\lambda(0, 1) = \lambda$, $\lambda(1, 0) = \mu$, $\phi(0) = 1$, and $\phi(1) = \phi$. Interpret 0 or 1 as indicating the absence or presence of a plant at a site. Thus sites remain vacant for periods of time which have mean λ^{-1}. The nominal lifetime is equal to the actual lifetime for a vacant period. Although the vacant period can be arbitrarily distributed it may be reasonable to suppose that it has an exponential distribution if, for example, plants appear through seeds settling at random from the atmosphere. After a plant appears it ages through its nominal lifetime, arbitrarily distributed with unit mean, at rate $\mu\phi^r$ while r of the neighbouring sites have plants at them. Depending on whether ϕ is greater or less than unity a plant shortens or lengthens the actual lifetime of its neighbours. The equilibrium distribution is

$$\pi(\mathbf{n}) = B\left(\frac{\lambda}{\mu}\right)^{\Sigma n_j}\phi^{-R} \tag{9.25}$$

where R is the number of neighbouring pairs of plants, and this distribution is insensitive to the form of the nominal lifetime distributions in the model.

Exercises 9.3

1. If the process $\mathbf{n}(t)$ has transition rates (9.15) show that the reversed process $\mathbf{n}(-t)$ has transition rates of the same form, but with $\lambda_j(n_j, m)$ replaced by $\alpha_j(m)\lambda_j(m, n_j)/\alpha_j(n_j)$.
2. Show that a closed reversible migration process $\mathbf{M}$, of the form introduced in Section 6.1, can be viewed as a reversible spatial process $\mathbf{n}$ in which

$$q(\mathbf{n}, T_j^m\mathbf{n}) = \lambda(n_j, m)\frac{\phi_{n_j}(M(n_j))}{M(n_j)}\psi_{n_m}(M(n_m))$$

   where the relationship between $\mathbf{M}$ and $\mathbf{n}$ is as in the preceding section. Observe that the process $\mathbf{n}$ is a reversible spatial process but does not have transition rates of the form (9.15).
3. Observe that the plant infection model leading to the equilibrium distribution (9.14) has transition rates of the form (9.15), with

$$\Phi(\mathbf{n}) = (1+\delta)^{-R(1)}$$

   and

$$\Phi_{G-j}(\mathbf{n}_{G-j}) = \Phi(T_j^1\mathbf{n}).$$

   Deduce that the equilibrium distribution (9.14) remains valid even when the duration of infection is arbitrarily distributed. Do the same for the plant infection model of Exercise 9.2.3. Observe that the process $\mathbf{n}(t)$, while not Markov, is reversible.
4. Show that the equilibrium distribution obtained for the power supply model of Exercise 9.2.4 is of the same form even when the period for which a source remains broken down is arbitrarily distributed.
5. Many of the models discussed in earlier chapters can readily be converted into spatial processes. Consider, for example, the model of a switching system described in Section 4.4. Show that if each of the K_1+K_2 lines is regarded as a site this model becomes a special case of the process described in this section.
6. A criticism of the plant birth and death model leading to the distribution (9.25) is that it is unlikely that the plants will be constrained to exist at a finite number of sites. Consider then the following model. Suppose that the points in time at which plants are born form a Poisson process and that when a plant is born its position is chosen at random from a uniform distribution over a fixed bounded region of the plane. Suppose, further, that a plant ages through its nominal lifetime at rate $\mu\phi^r$ where r is the number of plants within a distance d of it and $\phi \geq 1$. Approximate the process by the model of the preceding section, with the graph G taken as

a very fine grid of points covering the region and with two points of this grid defined as neighbours if they are within a distance d of each other. Use the approximation to show that conditional on there being N plants alive at a given point in time the probability they take up a given configuration in the region depends on that configuration only through R, the number of pairs of plants within a distance d of each other in the configuration. When $\phi<1$ the process does not have an equilibrium distribution; the expected number of plants in existence grows without limit. This difficulty was avoided, of course, when plants could exist at only a finite number of sites.

7. Elaborate migration processes can be constructed which have equilibrium distributions similar to those encountered in this chapter. In this exercise and the next we give some examples. Suppose the process $\mathbf{n}(t)$ has the following transition rates:

$$q(\mathbf{n}, T_{jk}\mathbf{n}) = \lambda_{jk}\frac{\Phi(\mathbf{n})}{\Phi(T_{j\cdot}\mathbf{n})}$$

$$q(\mathbf{n}, T_{j\cdot}\mathbf{n}) = \mu_j\frac{\Phi(\mathbf{n})}{\Phi(T_{j\cdot}\mathbf{n})}$$

$$q(\mathbf{n}, T_{\cdot k}\mathbf{n}) = \nu_k$$

Assuming an equilibrium distribution exists, show that it takes the form

$$\pi(\mathbf{n}) = B\frac{\prod_{j=1}^{J}\alpha_j^{n_j}}{\Phi(\mathbf{n})}$$

where $\alpha_1, \alpha_2, \ldots, \alpha_J$ is the solution of equations (2.9). Observe that we obtain the open migration process of Chapter 2 with

$$\Phi(\mathbf{n}) = \prod_{j=1}^{J}\prod_{r=1}^{n_j}\phi_j(r)$$

Observe also that the same procedure as used in the preceding section establishes that an individual's nominal lifetime in a colony can be arbitrarily distributed without affecting the equilibrium distribution $\pi(\mathbf{n})$.

8. Consider the following process in which each site can contain at most one particle. Let

$$q(\mathbf{n}, T_{jk}\mathbf{n}) = \lambda_{jk}\phi^{r}$$

$$q(\mathbf{n}, T_{j\cdot}\mathbf{n}) = \mu_j\phi^{r}$$

$$q(\mathbf{n}, T_{\cdot k}\mathbf{n}) = \nu_k$$

for j, k such that $n_j=1$, $n_k=0$, where r is the number of particles occupying sites neighbouring site j. Show that if $\alpha_1, \alpha_2, \ldots, \alpha_J$ satisfy equations (6.4) and (6.7) then the equilibrium distribution for the process is

$$\pi(\mathbf{n}) = B\phi^{-R} \prod_{j=1}^{J} \alpha_j^{n_j}$$

where R is the number of edges of the graph G which have occupied sites at both ends. Observe that $\pi(\mathbf{n})$ is a Markov field.

9.4 PARTIAL BALANCE

We have come across partial balance equations frequently in this work. In Chapter 3 we saw that quasi-reversibility was equivalent to a particular form of partial balance, and we have often found that models displaying partial balance possess an insensitivity property. In this section we shall investigate further the relationships between reversed processes, partial balance, and the phenomenon of insensitivity within the relatively simple setting provided by this chapter's definition of a spatial process.

The concept of partial balance was introduced in Chapter 1, where it was shown that some of the properties of a reversible process could be obtained from an assumption weaker than detailed balance. The following theorem summarizes Exercises 1.6.2, 1.6.3, 1.6.4, 1.7.7, and 1.7.8.

Theorem 9.5. *For a Markov process with transition rates $q(j, k)$, $j, k \in \mathcal{S}$, and equilibrium distribution $\pi(j)$, $j \in \mathcal{S}$, the following statements are equivalent:*

(i) *The distribution $\pi(j)$, $j \in \mathcal{S}$, satisfies the partial balance conditions*

$$\pi(j) \sum_{k \in \mathcal{A}} q(j, k) = \sum_{k \in \mathcal{A}} \pi(k) q(k, j) \qquad j \in \mathcal{A}$$

(ii) *If the process is truncated to the set $\mathcal{A}$ the equilibrium distribution of the truncated process is the conditional probability distribution*

$$\frac{\pi(j)}{\sum_{k \in \mathcal{A}} \pi(k)} \qquad j \in \mathcal{A}$$

(iii) *If the process is altered by changing the transition rate $q(j, k)$ to $cq(j, k)$ for $j, k \in \mathcal{A}$, where $c \neq 0$ or 1, then the resulting process has the unaltered equilibrium distribution $\pi(j)$, $j \in \mathcal{S}$.*

(iv) *If the process is altered by changing the transition rate $q(j, k)$ to $cq(j, k)$ for $j \in \mathcal{A}$, $k \in \mathcal{S} - \mathcal{A}$, where $c \neq 0$ or 1, then the equilibrium distribution of the resulting process takes the form*

$$\begin{array}{ll} B\pi(j) & j \in \mathcal{A} \\ Bc\pi(j) & j \in \mathcal{S} - \mathcal{A} \end{array}$$

(v) *The operations of time reversal and truncation to the set $\mathcal{A}$ commute.*

If in addition

$$\sum_{j \in \mathcal{A}} \sum_{k \in \mathcal{S}-\mathcal{A}} \pi(j)q(j, k)$$

is finite, then statements (i) *to* (v) *are equivalent to:*

(vi) *The Markov chain formed by observing the process at those instants in time just before it leaves the set* $\mathcal{A}$ *has the same equilibrium distribution as the Markov chain formed by observing the process at those instants in time just after it enters the set* $\mathcal{A}$.

When the Markov process has some additional structure the above results can usually be reformulated to make use of that structure. If the Markov process is a spatial process there are various truncations of the state space which take an intuitively appealing form. For example, single out a particular site j and suppose that the attributes of the other sites are held fixed at $\mathbf{n}_{G-j}$ with transitions involving changes at these sites forbidden. Under this truncation of the state space the attribute of site j, n_j, becomes a Markov process; the transition rate from n_j to m is $q(\mathbf{n}, T_j^m\mathbf{n})$ and thus depends on the frozen state of the rest of the system, $\mathbf{n}_{G-j}$, or at least on that part of it, $\mathbf{n}_{\partial j}$. Let $\pi(n_j; \mathbf{n}_{G-j})$, $n_j \in \mathcal{N}_j$, be the equilibrium distribution for this Markov process. Observe that a different truncation of the state space, and hence a different set of associated partial balance conditions, results from each choice of $\mathbf{n}_{G-j}$. Grouping these sets of partial balance conditions together we can obtain the following result.

Corollary 9.6. *For a spatial process with equilibrium distribution* $\pi(\mathbf{n})$, $\mathbf{n} \in \mathcal{S}$, *the following statements are equivalent:*

(i) *The distribution* $\pi(\mathbf{n})$, $\mathbf{n} \in \mathcal{S}$, *satisfies the partial balance equations*

$$\pi(\mathbf{n}) \sum_m q(\mathbf{n}, T_j^m\mathbf{n}) = \sum_m \pi(T_j^m\mathbf{n})q(T_j^m\mathbf{n}, \mathbf{n}) \qquad \mathbf{n} \in \mathcal{S} \qquad (9.26)$$

(ii) *The equilibrium distribution for the truncated process* n_j *obtained when sites other than* j *are frozen at* $\mathbf{n}_{G-j}$ *satisfies*

$$\pi(n_j; \mathbf{n}_{G-j}) = P(n_j \mid \mathbf{n}_{G-j}) \qquad \mathbf{n} \in \mathcal{S}$$

where $P(n_j \mid \mathbf{n}_{G-j})$ *is the conditional probability distribution (9.1).*

(iii) *If the process is altered by changing the transition rate* $q(\mathbf{n}, T_j^m\mathbf{n})$ *to* $cq(\mathbf{n}, T_j^m\mathbf{n})$ *for* $\mathbf{n} \in \mathcal{S}$, $m \in \mathcal{N}_j$, *where* $c \neq 0$ *or* 1, *then the resulting process has the unaltered equilibrium distribution* $\pi(\mathbf{n})$, $\mathbf{n} \in \mathcal{S}$.

(iv) *The operations of reversing time and freezing sites other than* j *at* $\mathbf{n}_{G-j}$ *commute, for all values of* $\mathbf{n}_{G-j}$.

The alteration to the process proposed in statement (ii) corresponds to a speeding up or slowing down of transitions altering the attribute of site j. Since the equilibrium distribution is unaffected by the value of the constant c the statement can be strengthened to allow a time-varying function $c(t)$. Indeed, a version of the statement can be formulated in which $c(t)$ is itself a stochastic process.

Since the transition rates $q(\mathbf{n}, T_j^m\mathbf{n})$ of the truncated process n_j do not depend on $\mathbf{n}_{G-j-\partial j}$ its equilibrium distribution $\pi(n_j; \mathbf{n}_{G-j})$ cannot depend on $\mathbf{n}_{G-j-\partial j}$ either. Hence from statement (ii) of Corollary 9.6 we obtain the following strengthening of Theorem 9.3.

Corollary 9.7. *If the equilibrium distribution of a spatial process satisfies the partial balance equations (9.26) for each $j \in G$ then it is a Markov field.*

From now on we shall not be much concerned with the spatial aspects of the process $\mathbf{n}(t)$ and the graph G may as well be the complete graph in which every pair of sites is connected by an edge. In this case the main feature of a spatial process $\mathbf{n}(t)$ is that only one of its components can change at a time.

Rather than freezing all sites except one suppose now that just one site is frozen. In particular suppose that the attribute of site j is frozen at n_j.

Corollary 9.8. *For a spatial process with equilibrium distribution $\pi(\mathbf{n})$, $\mathbf{n} \in \mathcal{S}$, the following statements are equivalent:*

(i) *The distribution $\pi(\mathbf{n})$, $\mathbf{n} \in \mathcal{S}$, satisfies the partial balance equations*

$$\pi(\mathbf{n}) \sum_m q(\mathbf{n}, T_j^m\mathbf{n}) = \sum_m \pi(T_j^m\mathbf{n}) q(T_j^m\mathbf{n}, \mathbf{n}) \qquad (9.27)$$

for the particular attribute n_j and for all values of $\mathbf{n}_{G-j}$.

(ii) *The equilibrium distribution for the truncated process $\mathbf{n}_{G-j}$ obtained when site j is frozen at n_j is the conditional probability distribution $P(\mathbf{n}_{G-j} \mid n_j)$ induced by the equilibrium distribution $\pi(\mathbf{n})$, $\mathbf{n} \in \mathcal{S}$.*

(iii) *If the process is altered by changing the transition rates $q(\mathbf{n}, T_j^m\mathbf{n})$ to $cq(\mathbf{n}, T_j^m\mathbf{n})$ for the particular attribute n_j and for all values of $\mathbf{n}_{G-j}$, m, where $c \neq 0$ or 1, then the equilibrium probability that the resulting process is in state $T_j^m\mathbf{n}$ takes the form*

$$\begin{array}{ll} B\pi(\mathbf{n}) & m = n_j \\ Bc\pi(T_j^m\mathbf{n}) & \text{otherwise} \end{array}$$

(iv) *The operations of reversing time and freezing site j at n_j commute.*

(v) *The equilibrium probability that the process is in state $\mathbf{n}$ given that the attribute of site j has just become n_j is the same as the equilibrium probability that the process is in state $\mathbf{n}$ given that the attribute of site j is about to change from n_j, for all values of $\mathbf{n}_{G-j}$.*

Proof. Let the set $\mathscr{A}$ be all states in $\mathscr{S}$ in which site j has the particular attribute n_j. Statement (i) is obtained from statement (i) of Theorem 9.5 by subtracting the partial balance conditions from the equilibrium equations. Statements (ii), (iii), and (iv) are just statements (ii), (iv), and (v) of Theorem 9.5. The probability flux out of the set $\mathscr{A}$ is finite since $\mathscr{S}$, and hence $\mathscr{A}$, is assumed to be finite. Thus statement (vi) of Theorem 9.5 applies, giving statement (v) of the present corollary.

Observe that the alteration to the process described in statement (iii) corresponds to a speeding up or slowing down of transitions ending the particular attribute n_j.

Of course freezing all sites other than one is not in principle very different from freezing just one site. Indeed, if we regard all sites other than one as forming a composite site the operations are identical. The important difference between Corollary 9.6 and Corollary 9.8 is that the partial balance equations (9.27) concern a particular attribute n_j while the partial balance equations (9.26) allow n_j to range over the set $\mathcal{N}_j$.

The consequences of partial balance described above are reminiscent of results obtained by other means in earlier chapters. There is, for example, a close relationship between statement (ii) of Theorem 9.5 and part (iii) of Theorem 3.12, or between statement (iv) of Theorem 9.5 and the product form obtained in Section 2.3. Statement (iv) of Corollary 9.6 gives an insight into why the reversed process obtained from a migration process takes such a simple form. These correspondences can be made precise by appropriately reformulating the models of earlier chapters, but this approach is generally too cumbersome to be useful.

We have not yet discussed the relationship between partial balance and the phenomenon of insensitivity, but the equivalence between statements (i) and (ii) of Corollary 9.6 suggests the following very rough line of argument. If the nominal lifetimes of the attributes at site j are arbitrarily rather than exponentially distributed then this will not affect the equilibrium distribution of the truncated process n_j. Hence it should not affect the equilibrium distribution $\pi(\mathbf{n})$ of the process $\mathbf{n}(t)$. We shall now make this argument more precise.

Let $\mathbf{x}=(x_1, x_2, \ldots, x_J)$ be a spatial process with state space $\mathscr{X}_1 \times \mathscr{X}_2 \times \cdots \times \mathscr{X}_J$, and let $\pi(x_j; \mathbf{x}_{G-j})$, $x_j \in \mathscr{X}_j$, be the equilibrium distribution for the truncated process x_j obtained when sites other than j are frozen at $\mathbf{x}_{G-j}$. Thus

$$\pi(x_j; \mathbf{x}_{G-j}) \sum_y q(\mathbf{x}, T_j^y\mathbf{x}) = \sum_y \pi(y; \mathbf{x}_{G-j}) q(T_j^y\mathbf{x}, \mathbf{x}) \tag{9.28}$$

Suppose that for each $j \in G$ there is a function

$$f_j: \mathscr{X}_j \to \mathcal{N}_j$$

Let $n_j = f_j(x_j)$ and write $\mathbf{n} = (n_1, n_2, \ldots, n_J)$ for $f(\mathbf{x}) = (f_1(x_1), f_2(x_2), \ldots, f_J(x_J))$. Thus n_j is a less detailed description than x_j of the attribute of site j. We can imagine that the dependence between site j and sites other than j is through n_j rather than x_j. To formalize this idea we shall make two assumptions about the truncated process x_j. The first assumption is that the transition rates $q(\mathbf{x}, T_j^y\mathbf{x})$ depend on $\mathbf{x}_{G-j}$ only through $\mathbf{n}_{G-j}$. Thus $\pi(x_j; \mathbf{x}_{G-j})$ can depend on $\mathbf{x}_{G-j}$ only through $\mathbf{n}_{G-j}$ and so can be written $\pi(x_j; \mathbf{n}_{G-j})$. Let

$$\pi(n_j; \mathbf{n}_{G-j}) = \sum_{x_j : f_j(x_j) = n_j} \pi(x_j; \mathbf{n}_{G-j})$$

The second assumption is that $\pi(x_j; \mathbf{n}_{G-j})$ can be written in the form

$$\pi(x_j; \mathbf{n}_{G-j}) = \pi(n_j; \mathbf{n}_{G-j}) P_j(x_j \mid n_j) \tag{9.29}$$

This assumption is best regarded as a condition on the function f_j; put in statistical terms relation (9.29) asserts that the statistic $n_j = f_j(x_j)$ obtained from the data x_j is sufficient for the parameter $\mathbf{n}_{G-j}$. Put another way the assumption is that conditional on n_j the attribute x_j has a distribution $P_j(x_j \mid n_j)$ which does not depend on $\mathbf{n}_{G-j}$. Both assumptions together allow us to write

$$\pi(x_j; \mathbf{x}_{G-j}) = \pi(n_j; \mathbf{n}_{G-j}) P_j(x_j \mid n_j) \tag{9.30}$$

where $\mathbf{n} = f(\mathbf{x})$.

If we had to choose transition rates for a spatial process so that it resembled the process $\mathbf{n} = f(\mathbf{x})$ how would we do it? One way would be to set

$$q(\mathbf{n}, T_j^m\mathbf{n}) = \sum_{x_j : f_j(x_j) = n_j} P_j(x_j \mid n_j) \sum_{y : f_j(y) = m} q(\mathbf{x}, T_j^y\mathbf{x}) \tag{9.31}$$

Observe that if the truncated process x_j is in equilibrium the right-hand side of equation (9.31) is the probability intensity that $f_j(x_j)$ changes to m given that it starts equal to n_j.

Theorem 9.9. *Suppose a spatial process $\mathbf{x}(t)$ is such that the truncated process x_j satisfies both the above assumptions for each $j \in G$. Let the rates $q(\mathbf{n}, T_j^m\mathbf{n})$ be defined by equation (9.31). If there exists a distribution $\pi(\mathbf{n})$ satisfying the partial balance equations (9.26) for each $j \in G$ then the equilibrium distribution for the process $\mathbf{x}(t)$ is*

$$\pi(\mathbf{x}) = \pi(\mathbf{n}) \prod_{j=1}^{J} P_j(x_j \mid n_j) \tag{9.32}$$

and it satisfies the partial balance equations

$$\pi(\mathbf{x}) \sum_{y \in \mathscr{X}_j} q(\mathbf{x}, T_j^y\mathbf{x}) = \sum_{y \in \mathscr{X}_j} \pi(T_j^y\mathbf{x}) q(T_j^y\mathbf{x}, \mathbf{x})$$

for each $j \in G$ and for all $\mathbf{x}$.

Proof. Substituting the form (9.30) into equations (9.28) we obtain

$$\pi(n_j; \mathbf{n}_{G-j})P_j(x_j \mid n_j) \sum_{m \in \mathcal{N}_j} \sum_{y: f_j(y)=m} q(\mathbf{x}, T_j^y \mathbf{x})$$
$$= \sum_{m \in \mathcal{N}_j} \sum_{y: f_j(y)=m} \pi(m; \mathbf{n}_{G-j})P_j(y \mid m)q(T_j^y \mathbf{x}, \mathbf{x}) \qquad (9.33)$$

If these equations are summed over x_j such $f_j(x_j) = n_j$ they reduce to

$$\pi(n_j; \mathbf{n}_{G-j}) \sum_m q(\mathbf{n}, T_j^m \mathbf{n}) = \sum_m \pi(m; \mathbf{n}_{G-j})q(T_j^m \mathbf{n}, \mathbf{n}) \qquad (9.34)$$

using equation (9.31). Since $\pi(\mathbf{n})$ satisfies the partial balance equations the distribution satisfying equations (9.34) is

$$\pi(n_j; \mathbf{n}_{G-j}) = \frac{\pi(\mathbf{n})}{\sum_{m \in \mathcal{N}_j} \pi(T_j^m \mathbf{n})}$$

Substituting this back into equations (9.33) confirms that the proposed equilibrium distribution $\pi(\mathbf{x})$ satisfies the partial balance equations for the process $\mathbf{x}(t)$. Hence $\pi(\mathbf{x})$ is indeed the equilibrium distribution for the process $\mathbf{x}(t)$.

Consider now a spatial process $\mathbf{n}$ with transition rates $q(\mathbf{n}, T_j^m \mathbf{n})$. Consider in particular the period of time for which a site's attribute remains unchanged. We can imagine that after the attribute of site j becomes n_j the attribute has a nominal lifetime exponentially distributed with unit mean which it ages through at rate

$$\sum_m q(\mathbf{n}, T_j^m \mathbf{n}) \qquad (9.35)$$

and that when the attribute's lifetime ends site j takes on attribute n with probability

$$\frac{q(\mathbf{n}, T_j^n \mathbf{n})}{\sum_m q(\mathbf{n}, T_j^m \mathbf{n})} \qquad (9.36)$$

Clearly this description of the evolution of the system is consistent with the transition rates $q(\mathbf{n}, T_j^m \mathbf{n})$. Now suppose that an attribute's nominal lifetime has some arbitrary distribution with unit mean, where this distribution may vary from attribute to attribute and from site to site. The process $\mathbf{n}(t)$ will no longer be a Markov process, but if the partial balance equations (9.26) are satisfied its equilibrium distribution will still be given by $\pi(\mathbf{n})$. We shall establish this fact for the case where nominal lifetimes are distributed as mixtures of gamma distributions using the method of stages.

Suppose that when site j takes on attribute m it in fact takes on a finer attribute, (m, z), with probability $p_j(m, z)$, where z belongs to a countable set

$\mathscr{Z}$ and $\sum_z p_j(m, z) = 1$ for each j and z. Suppose that the nominal lifetime of the finer attribute is made up of $w_j(m, z)$ independent stages, each exponentially distributed with mean $d_j(m, z)$. This framework allows the nominal lifetime of attribute m at site j to have any distribution which can be expressed as a mixture of gamma distributions. Set

$$\sum_z p_j(m, z)w_j(m, z)d_j(m, z) = 1 \tag{9.37}$$

so that nominal lifetimes have unit mean. Use $x_j = (n_j, z_j, u_j)$ to describe site j where (n_j, z_j) is the finer attribute of the site and $u_j (1 \le u_j \le w_j(n_j, z_j))$ is the stage currently in progress. Observe that the description $\mathbf{x} = (x_1, x_2, \ldots, x_J)$ is detailed enough to be a Markov process. Consider now the reduction $\mathbf{n} = f(\mathbf{x})$ given by $n_j = f_j(x_j)$, $j \in G$. The transition rates of the truncated process x_j depend on $\mathbf{x}_{G-j}$ only through $\mathbf{n}_{G-j}$, and hence we can write

$$\pi(x_j; \mathbf{x}_{G-j}) = \pi(x_j; \mathbf{n}_{G-j})$$

Considering the truncated process x_j more closely it is apparent that although $\mathbf{n}_{G-j}$ affects the *rate* at which transition occur it does not affect *which* transitions occur except possibly when n_j changes. If we observe the truncated process x_j only while n_j takes the value m the proportion of time for which (z_j, u_j) takes the value (z, u) is $p_j(m, z)d_j(m, z)$ and is unaffected by $\mathbf{n}_{G-j}$. Hence we can write

$$\pi(x_j; \mathbf{n}_{G-j}) = \pi(n_j; \mathbf{n}_{G-j})P_j(x_j \mid n_j)$$

where

$$P_j(x_j \mid n_j) = P_j(z_j, u_j \mid n_j) = p_j(n_j, z_j)d_j(n_j, z_j) \qquad z_j \in \mathscr{Z}, \quad 1 \le u_j \le w_j(n_j, z_j)$$

Equation (9.37) ensures that the distribution $P_j(x_j \mid n_j)$ sums to unity. It is readily checked that the rates $q(\mathbf{n}, T_j^m\mathbf{n})$ are consistent with the definition (9.31). We can thus use Theorem 9.9 to deduce that if $\pi(\mathbf{n})$ satisfies the partial balance equations (9.26) then the equilibrium distribution for the process $\mathbf{x}$ takes the form (9.32). Hence the process $\mathbf{n}$ has the equilibrium distribution $\pi(\mathbf{n})$ and is insensitive to the exact form of the nominal lifetime distributions provided they arise as mixtures of gamma distributions. The restriction to mixtures of gamma distributions can be removed, but we shall not discuss this.

There are stronger forms of insensitivity than that just described, and we shall briefly describe one. Suppose the transition rates $q(\mathbf{n}, T_j^m\mathbf{n})$ allow a solution to the partial balance equations (9.26) and further suppose the ratio (9.36) does not depend upon $\mathbf{n}_{G-j}$, as, for example, in the general spatial process described in the last section. We can then allow the next attribute to be taken on by site j and that attribute's nominal lifetime to depend upon the previous sequence of attributes taken by site j and upon their nominal lifetimes: if the process is stationary the equilibrium distribution for $\mathbf{n}$ will be insensitive to these dependencies. The reader will observe the parallel with the pattern of dependencies allowed in closed queueing networks. We shall not prove the result, but the idea is to extend the description x_j to include a summary of the history of site j. The condition on the ratio (9.36) ensures that the jump chain of the truncated process x_j is independent of $\mathbf{n}_{G-j}$, and this in turn shows that $\mathbf{n}_{G-j}$ and x_j are independent, conditional on n_j. The result is not really surprising in view of Corollary 9.6; the equilibrium distribution for the truncated process n_j will be determined by the proportion of time for which site j takes each attribute, no matter how complicated the pattern of dependencies.

In the above discussion it was assumed that the equilibrium distribution of the spatial process satisfied all the partial balance equations (9.26). If the equilibrium distribution satisfies the smaller set of partial balance equations described in statement (i) of Corollary 9.8 then it is possible to show that the nominal lifetime of the particular attribute n_j can be arbitrarily distributed with unit mean without it affecting the equilibrium distribution $\pi(\mathbf{n})$ (Exercise 9.4.1); this might be intuitively expected in view of the equivalence between statements (i) and (ii) of Corollary 9.8. We shall now prove a converse to this result.

Theorem 9.10. *If in a spatial process the nominal lifetime of a particular attribute, n_j, can have any distribution with unit mean without it affecting the equilibrium distribution $\pi(\mathbf{n})$, $\mathbf{n} \in \mathcal{S}$, then*

$$\pi(\mathbf{n}) \sum_m q(\mathbf{n}, T_j^m \mathbf{n}) = \sum_m \pi(T_j^m \mathbf{n}) q(T_j^m \mathbf{n}, \mathbf{n}) \tag{9.38}$$

for the particular attribute n_j and for all values of $\mathbf{n}_{G-j}$.

Proof. The equilibrium equations for the case where the nominal lifetime of attribute n_j is exponentially distributed show that

$$\pi(\mathbf{n}) \sum_i \sum_m q(\mathbf{n}, T_i^m \mathbf{n}) = \sum_i \sum_m \pi(T_i^m \mathbf{n}) q(T_i^m \mathbf{n}, \mathbf{n}) \tag{9.39}$$

Now let the nominal lifetime of attribute n_j be a mixture of two exponential random variables. Suppose that when site j takes on attribute n_j the nominal lifetime of the attribute is exponentially distributed with mean either a_1 or a_2, each possibility being equally likely. To ensure that the mean nominal lifetime is unity, set $a_1 + a_2 = 2$. The process $\mathbf{n}$ can be rendered Markov if when site j has attribute n_j we append an indication of which exponential random variable has been chosen for the attribute's nominal lifetime. Write $\pi(\mathbf{n}, 1)$, $\pi(\mathbf{n}, 2)$, $\pi(T_j^m \mathbf{n})$ for the equilibrium distribution of the resulting process. The equilibrium equations for this process show that

$$\pi(\mathbf{n}, 1)\left(a_1^{-1}\sum_m q(\mathbf{n}, T_j^m\mathbf{n})+\sum_{i\neq j}\sum_m q(\mathbf{n}, T_i^m\mathbf{n})\right)$$
$$=\tfrac{1}{2}\sum_m \pi(T_j^m\mathbf{n})q(T_j^m\mathbf{n}, \mathbf{n})+\sum_{i\neq j}\sum_m \pi(T_i^m\mathbf{n}, 1)q(T_i^m\mathbf{n}, \mathbf{n}) \qquad (9.40)$$

and

$$\pi(\mathbf{n}, 2)\left(a_2^{-1}\sum_m q(\mathbf{n}, T_j^m\mathbf{n})+\sum_{i\neq j}\sum_m q(\mathbf{n}, T_i^m\mathbf{n})\right)$$
$$=\tfrac{1}{2}\sum_m \pi(T_j^m\mathbf{n})q(T_j^m\mathbf{n}, \mathbf{n})+\sum_{i\neq j}\sum_m \pi(T_i^m\mathbf{n}, 2)q(T_i^m\mathbf{n}, \mathbf{n}) \qquad (9.41)$$

Now

$$\pi(\mathbf{n}, 1)+\pi(\mathbf{n}, 2)=\pi(\mathbf{n}) \qquad (9.42)$$

and

$$\pi(T_i^m\mathbf{n}, 1)+\pi(T_i^m\mathbf{n}, 2)=\pi(T_i^m\mathbf{n})$$

so if equations (9.40) and (9.41) are added together and the result compared with equation (9.39) we obtain

$$\pi(\mathbf{n}, 1)a_1^{-1}+\pi(\mathbf{n}, 2)a_2^{-1}=\pi(\mathbf{n}) \qquad (9.43)$$

Solving equations (9.42) and (9.43) gives

$$\pi(\mathbf{n}, 1)=\tfrac{1}{2}a_1\pi(\mathbf{n})$$
$$\pi(\mathbf{n}, 2)=\tfrac{1}{2}a_2\pi(\mathbf{n})$$

Substituting for $\pi(\mathbf{n}, 1)$, $\pi(T_i^m\mathbf{n}, 1)$ in equation (9.40) we obtain

$$\pi(\mathbf{n})\left(\tfrac{1}{2}\sum_m q(\mathbf{n}, T_j^m\mathbf{n})+a_1\sum_{i\neq j}\sum_m q(\mathbf{n}, T_i^m\mathbf{n})\right)$$
$$=\tfrac{1}{2}\sum_m \pi(T_j^m\mathbf{n})q(T_j^m\mathbf{n}, \mathbf{n})+a_1\sum_{i\neq j}\sum_m \pi(T_i^m\mathbf{n})q(T_i^m\mathbf{n}, \mathbf{n})$$

and since this equation holds for more than one value of a_1 we can deduce the result stated.

Exercises 9.4

1. If equations (9.38) hold for the particular attribute n_j and for all values of $\mathbf{n}_{G-j}$ then the equilibrium distribution $\pi(\mathbf{n})$ is insensitive to the form of the nominal lifetime distribution of attribute n_j, even if all the partial balance equations (9.26) do not hold. Establish this either directly or by appending a marker site to the system in such a way that all the partial balance equations for the extra site hold.

2. Section 6.1 contained the assertion that the equilibrium distribution $\pi(\mathbf{n})$ of a closed reversible migration process is insensitive to the form of the nominal lifetime distributions. In fact the nominal lifetimes of an individual at successive colonies can even be dependent on each other without it affecting the equilibrium distribution $\pi(\mathbf{n})$. Establish this in the case where the process can be represented by a stationary Markov process with a countable state space. Observe that since the ratio (9.36) depends on $\mathbf{n}_{G-j}$ it is not easy to formulate a version of the process in which there exist arbitrary dependencies between the route of an individual and the nominal lifetimes along that route.
3. Consider the following model of competition between two species. Let n_i be the population size of species i, for $i=1, 2$. The probability intensity that a new individual of species 1 is born is n_1+1 and that an individual of species 1 dies is $n_1 \exp(-\alpha+\gamma(n_1-1)+\delta n_2)$, $\gamma, \delta>0$. The birth and death rates for species 2 are the same, with n_1 and n_2 interchanged. Observe that the form of the birth rate prevents a species from becoming extinct and corresponds to immigration at unit rate. The probability intensity that a given individual dies is increased by a factor e^{γ} by each other existing individual of the same species and by a factor e^{δ} by each other existing individual of the other species. Show that in equilibrium

$$\pi(n_1, n_2)=B \exp\left\{\alpha(n_1+n_2)-\frac{\gamma}{2}[n_1(n_1-1)+n_2(n_2-1)]-\delta n_1 n_2\right\}$$

The most probable states (n_1, n_2) fall within regions whose position depends upon the parameters α, γ, and δ. Sketch the quadratic contours along which the function $\pi(n_1, n_2)$ takes a constant value and obtain Fig. 9.2. Show that the equilibrium distribution $\pi(n_1, n_2)$ is unaffected if individuals have arbitrarily distributed nominal lifetimes.
4. Suppose the equilibrium distribution $\pi(\mathbf{n})$, $\mathbf{n}\in\mathcal{S}$, of a spatial process satisfies the partial balance equations (9.26) for each $j\in G$. Suppose now that the process is amended so that the nominal lifetimes of attributes are allowed to be arbitrarily distributed with *any* mean. Let $\alpha_j(n_j)$ be the mean nominal lifetime of attribute n_j at site j. Use Corollary 9.8 to show that the equilibrium probability that $\mathbf{n}$ describes the state of the amended process is of the form

$$B\pi(\mathbf{n})\prod_j \alpha_j(n_j) \qquad \mathbf{n}\in\mathcal{S}$$

5. Exercise 9.4.1 and Theorem 9.10 characterize a certain form of insensitivity in terms of partial balance equations, within the framework provided by the definition of a spatial process. Use Exercise 4.6.9 to define a spatial process exhibiting an alternative form of insensitivity.
6. Statement (vi) of Theorem 9.5 provides an alternative explanation of the connection between partial balance and insensitivity. Suppose that for a

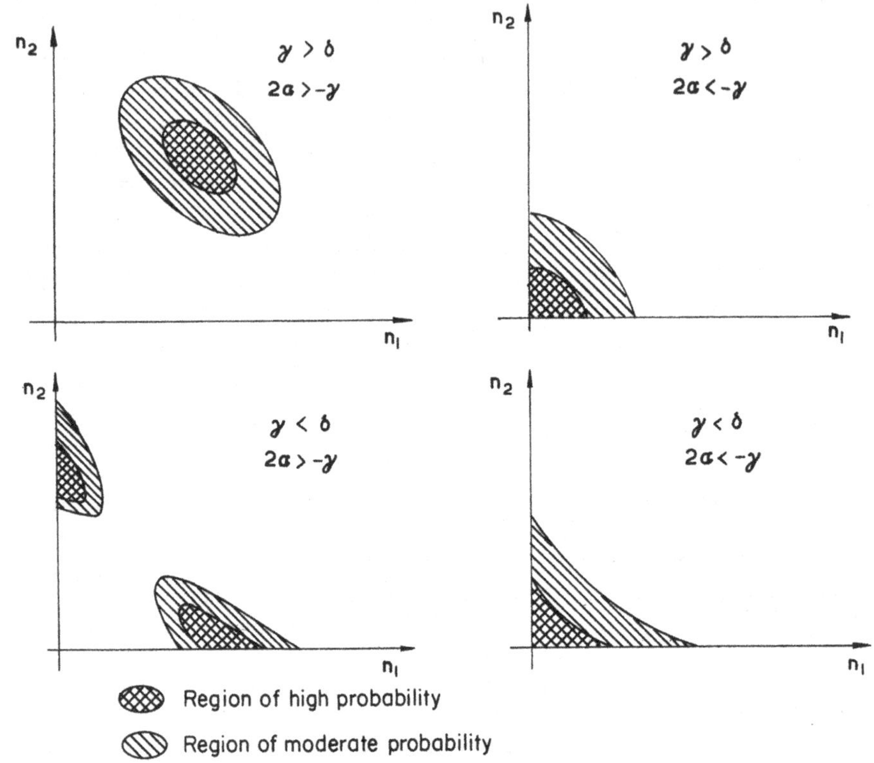

Fig. 9.2 A competition model

Markov process $X(t)$ with transition rates $q(j, k)$, $j, k \in \mathscr{S}$, and equilibrium distribution $\pi(j)$, $j \in \mathscr{S}$, statement (i) of Theorem 9.5 holds. Suppose also that $\sum_{k \in \mathscr{S}-\mathscr{A}} q(j, k)$ is the same for each $j \in \mathscr{A}$, equal to $q(\mathscr{A})$ say, and that $\mathscr{A}$ is finite; hence statement (vi) of Theorem 9.5 is true. Show that in equilibrium the distribution over states at an instant in time when the process has just entered the set $\mathscr{A}$ and the distribution over states at an instant in time when the process is just about to leave the set $\mathscr{A}$ are both identical to the equilibrium distribution of the truncated process referred to in statement (ii) of Theorem 9.5. Consider now a realization of the process $X(t)$. The process will alternate between the set $\mathscr{A}$ and the set $\mathscr{S}-\mathscr{A}$, and the periods spent in the set $\mathscr{A}$ will be exponentially distributed with parameter $q(\mathscr{A})$. Because of the identity between the three distributions referred to above the behaviour of the process during a period in the set $\mathscr{A}$ is as if it were generated by the following procedure. When the process enters the set $\mathscr{A}$ choose a random variable τ, exponentially distributed with parameter $q(\mathscr{A})$. For a period of length τ allow the process to evolve in accordance with the transition rates $q(j, k)$,

$j, k \in \mathscr{A}$, of the truncated process. At the end of this period eject the process from the set $\mathscr{A}$: if the process is in state $j \in \mathscr{A}$ move it to state $k \in \mathscr{S} - \mathscr{A}$ with probability $q(j, k)/q(\mathscr{A})$. The advantage of viewing the process in this way is that it becomes natural to allow the periods spent in the set $\mathscr{A}$ to be arbitrary rather than exponential random variables. The proportion of time the process spends in the set $\mathscr{A}$ will then be determined by the overall mean of these periods, and of the time spent in the set $\mathscr{A}$ a proportion $\pi(j)/\sum_{k \in \mathscr{A}} \pi(k)$ will be spent in state j.

References

1. MARKOV PROCESSES AND REVERSIBILITY

For a fuller introduction to Markov processes see Feller (1968) and Cox and Miller (1965); a thorough account of the general theory is given by Chung (1967). Time-reversed Markov chains were first considered by Kolmogorov (1936); this paper contains his criteria for reversibility, which were later extended by Reich (1957) and Kendall (1960). For a discussion of reversible diffusion processes see Kent (1978). Good introductions to queueing theory are Cox and Smith (1961), Cooper (1972), Gross and Harris (1974), Kleinrock (1975), and Koenig and Stoyan (1976); Little's result is proved in Ross (1970). For a further discussion of the Ehrenfest model see Kac (1959), Bartlett (1962), and Ruben (1964). Theorem 1.6 is taken from Moran (1961); the result is used by Rényi (1970) and Penrose (1970) to establish the convergence to a limiting distribution of a Markov chain. The electrical network described in Section 1.4 is used by Keilson (1965) in his review of the transient behaviour of reversible processes. Exercise 1.2.6 and the concept of dynamic reversibility are due to Whittle (1975). For uses of dynamic reversibility in spectral analysis see Anderson and Kailath (1979). Reversible null recurrent Markov processes are discussed by Kendall (1975); in this setting Exercises 1.2.7 and 1.7.8 take a more natural form.

2. MIGRATION PROCESSES

Theorem 2.1 was originally proved by Burke (1956). The proof given here based on reversibility is due to Reich (1957). The queue considered in Exercise 2.1.1 was described by Boes (1969); the method used in part (i) of this exercise and in the telephone exchange model of Section 2.1 is due to Daley (1976). A series of simple queues was first considered by Jackson (1954). Theorem 2.2 was originally proved by Reich (1957, 1963, 1965); the proof given here is due to Burke (1972), who also established the results contained in Exercises 2.2.4, 2.2.5, and 2.2.6. For the solution to Exercise 2.2.4 see Burke (1968). Theorems 2.3 and 2.4 come from the important paper of Jackson (1963). The phenomenon of partial balance was observed by Whittle (1967, 1968). The examples in Section 2.3 are based on work by Taylor and Jackson (1954) and Koenigsberg (1958). Exercise 2.3.5 is a development of Gordon and Newell (1967). The reader interested in the computational methods touched on in Exercises 2.3.6, 2.3.7, and 2.3.8

should see Reiser (1977). Corollary 2.6 is due to Muntz (1972) and Theorem 2.5 to Kelly (1975a). The family size process is described by Kendall (1975) and the optimal allocation result is due to Kleinrock (1964). The converse to Exercise 2.4.3 has been established by Melamed (1979) for the case where each colony is a single-server queue; see also Brémaud (1978). Exercise 2.4.5 comes from Whittle (1968).

3. QUEUEING NETWORKS

The important paper of Muntz (1972) established the link between Poisson preserving queues and product form solutions for networks. This link is the basis of the treatment of queueing networks given in this chapter. Baskett, Chandy, Muntz, and Palacios (1975) exhibited the equilibrium distribution for networks in which the queues take certain forms, these forms including most of the special cases discussed in Sections 3.1 and 3.3. The presentation given in Sections 3.1 and 3.3 is based on Kelly (1976b). The means whereby an arbitrary distribution is approximated by a mixture of gamma distributions is known as the method of stages (Cox and Miller, 1965). The method is particularly useful in Section 3.3 because of the insensitivity of the results to the form of the distributions, and hence to the complexity of the approximation. Barbour (1976) establishes the validity of the limiting procedure used in Section 3.3 whereby systems involving arbitrary distributions are approximated by systems involving mixtures of gamma distributions. Results for systems involving arbitrary distributions can be obtained directly: see Takács (1969), Shanbhag and Tambouratzis (1973), and Kelly (1976d). The insensitivity of symmetric queues can be deduced from the results of Koenig, Matthes, and Nawrotzki (1967); see also Chandy, Howard, and Towsley (1977). The phenomenon described in part (iii) of Theorem 3.12 was observed in a class of teletraffic models by Cohen (1957) and Descloux (1967). The expression of the virtual waiting time in an $M/G/1$ queue as a geometric mixture was first given by Beneš (1957); Exercise 3.3.7 supplies an explanation of the fact. Exercise 3.3.10 is taken from Kelly (1975a). The results contained in Exercise 3.3.11 were obtained by Shanbhag and Tambouratzis (1973), Mecke (1975), and Oakes (1976). Exercise 3.4.6 is due to Cohen (1977) and Schassberger (1978b). Exercise 3.4.8 is based on a result of Mar'yanovitch discussed by Gnedenko and Kovalenko (1968). The functional form (3.26) is suggested by the work of Kovalenko (1962). Exercise 3.5.8 is based on an example of Koenig and Jansen (1974).

Recent reviews of work on queueing networks are Disney (1975), Gelenbe and Muntz (1976), and Lemoine (1977). Sections on queueing networks are included in the textbook of Kleinrock (1975) and the monograph of Curtois (1977).

4. EXAMPLES OF QUEUEING NETWORKS

The approach to communication networks described in Section 4.1 was pioneered by Kleinrock (1964, 1976). For an introduction to the area of machine interference see Cox and Smith (1961). Muntz (1975) reviews queueing models of computer systems. Exercises 4.2.5 and 4.3.4 are based on an idea of Schassberger. Syski (1960), Beněs (1965), and Cooper (1972) discuss many interesting teletraffic models. The model of a telephone exchange with unreliable lines generalizes work by Mar'yanovitch discussed in Gnedenko and Kovalenko (1968). The model of a switching system extends a result of Buchner and Neal (1971). Exercise 4.4.9 describes a result of Cohen (1957) and Exercise 4.4.10 a result of Koenig (1965). Cooper (1972) gives some of the background to the interesting phenomenon mentioned in Exercise 4.4.11. Matis and Hartley (1971) discuss compartmental models in biology, Chiang (1968) birth–illness–death models, and Bartholomew (1973) manpower systems. Theorem 4.2 and Exercise 4.5.4 are taken from Kingman (1969). The transient analysis of compartmental models in which the arrival rate depends on the number in the system is difficult (see Milch, 1968). Takács (1962) discusses the transient behaviour of certain queues. Exercise 4.5.6 (and its solution) come from Oakes (1977). The problem with viewing road traffic flow as a series of queues is that the queues may coalesce: see Miller (1961) and Gipps (1977). The generalization of a type I counter is mentioned in Oakes (1976). For the more usual approach to electronic counters, via renewal theory, see Cox (1962). The result for a repair shop is due to Yaroshenko and is discussed in Gnedenko and Kovalenko (1968). Exercise 4.6.9 is based on Wolff and Wrightson (1976). The applications mentioned in Exercises 4.6.11 and 4.6.12 are discussed by Maher and Cabrera (1975) and Koenigsberg and Lam (1976).

5. ELECTRICAL ANALOGUES

The connection between reversible random walks and electrical networks described in Section 5.1 has been discussed by Nash-Williams (1959) and Kemeny, Snell, and Knapp (1976). The flow model of Section 5.2 was introduced by Kingman (1969) and the invasion model of Section 5.3 by Clifford and Sudbury (1973); both of these models were considered further by Kelly (1976c).

6. REVERSIBLE MIGRATION PROCESSES

The reversible migration processes introduced in Section 6.1 come from the important paper of Kingman (1969). The models of social grouping behaviour discussed in Section 6.2 are taken from Cohen (1972); see also

Cohen (1971). The closed flow model leading to the equilibrium distribution (6.14) is due to Spitzer (1970). Exercise 6.3.2 is based on an idea of White (1970).

7. POPULATION GENETICS MODELS

The neutral allele model discussed is just one of the many stochastic models of use in population genetics; others are described by Ewens (1969) and Crow and Kimura (1970). For an account of the varying emphasis in current theories of evolution see Crow (1972), Lewontin (1974), and Kimura (1976). The model introduced in Section 7.1 is based on the work of Moran (1958); its equilibrium distribution was obtained by Trajstman (1974). Theorem 7.1 comes from Kelly (1977); Ewens (1972) and Watterson (1974) had previously established the result for samples from large populations. The results contained in Exercises 7.1.4, 7.1.8, and 7.1.11 are due to Ewens (1972). Exercises 7.1.6 and 7.1.7 are based on Kingman (1978) and Exercise 7.1.10 on Kelly (1976a). It was Watterson (1976) who first realized that reversibility could help answer questions about the age of an allele. Section 7.2 is taken mainly from Watterson and Guess (1977) and Kelly (1977). Fixation times are discussed by Guess and Ewens (1972); the results of Section 7.3 come from Kelly (1977). Exercise 7.3.2 is due to Watterson (1961); the extension contained in Exercise 7.3.3 is a variant of a result of Kimura and Ohta (1969). For a fuller discussion of the topic touched on in Exercise 7.3.5 see Watterson (1975). Another application of time reversed processes in genetics is described by Seneta (1965).

8. CLUSTERING PROCESSES

Early attempts to model social grouping behaviour were made by Coleman and James (1961), White (1962), and Goodman (1964). The basic model of Section 8.2 is due to Whittle (1965a). Morgan (1976) established Lemma 8.3 and discussed some of the examples given in Section 8.3. Expression (8.21) was first obtained through the classic approach to polymer statistics of Flory (1953) and Stockmayer (1943). Watson (1958), Gordon (1962), and Good (1963) developed the branching process approach touched on in Exercise 8.4.7. The development presented in Section 8.4 is a special case of the powerful and elegant treatment of polymerization processes due to Whittle (1965a, 1965b, 1972).

9. SPATIAL PROCESSES

Theorem 9.2 was first established by Brook (1964); Grimmett (1973) obtained the explicit representation given in Exercise 9.1.1. See Moussouris

(1974) for the counterexamples referred to in Exercise 9.2.2. Methods for the statistical analysis of data arising from Markov fields have been developed by Besag (1974a); see also Kingman (1975), Ripley (1977), Bartlett (1978), and the papers in Tweedie (1978). Theorem 9.3 is due to Spitzer (1971) and Besag (1974b). Much of the work on Markov fields is motivated by applications in statistical mechanics; expositions from this point of view are Preston (1974) and Spitzer (1974). Section 9.3 is based on Kelly (1975b). The step from gamma distributions to arbitrary distributions is made possible by the work of Barbour (1976). The reader interested in the model developed in Exercise 9.3.6 should see Preston (1975), Kelly and Ripley (1976), and Ripley and Kelly (1977). Matthes and Koenig have done important work on the phenomenon of insensitivity and have shown that for a certain fairly general class of systems insensitivity is characterized by a form of partial balance (see the appendix to Gnedenko and Kowalenko, 1971, written by Koenig, Matthes, and Nawrotzki, or the introduction to this work written by Schassberger, 1977). Theorems 9.9 and 9.10 are inspired by this work. It is possible to describe a network of symmetric queues as a spatial process (cf. Schassberger, 1978a). Exercise 9.4.3 is taken from Whittle (1975).

Anderson, B. D. O., and Kailath, T. (1979). Forwards and backwards models for finite-state Markov processes. *Adv. Appl. Prob.*, **11,** 118–133.

Barbour, A. D. (1976). Networks of queues and the method of stages. *Adv. Appl. Prob.*, **8,** 584–591.

Bartholomew, D. J. (1973). *Stochastic Models for Social Processes*, 2nd ed., Wiley, London.

Bartlett, M. S. (1962). Irreversibility and statistical theory. In *Essays on Probability and Statistics*, Methuen, London.

Bartlett, M. S. (1978). *An Introduction to Stochastic Processes*, 3rd ed., Cambridge University Press.

Baskett, F., Chandy, M., Muntz, R., and Palacios, J. (1975). Open, closed and mixed networks of queues with different classes of customers. *J.A.C.M.*, **22,** 248–260.

Beněs, V. E. (1957). On queues with Poisson arrivals. *Ann. Math. Stat.*, **28,** 670–677.

Beněs, V. E. (1965). *Mathematical Theory of Connecting Networks and Telephone Traffic*, Academic Press, London.

Besag, J. (1974a). Spatial interaction and the statistical analysis of lattice systems. *J. Roy. Statist. Soc.*, B**36,** 192–236.

Besag, J. (1974b). On spatial-temporal models and Markov fields. *Transactions of the Seventh Prague Conference on Information Theory, Statistical Decision Functions, and Random Processes, and of the* 1974 *European Meeting of Statisticians*, Czechoslovak Academy of Sciences.

Boes, D. C. (1969). Note on the output of a queueing system. *J. Appl. Prob.*, **6,** 459–461.

Brémaud, P. (1978). Streams of a *M/M/*1 feedback queue in statistical equilibrium. *Z. Wahrscheinlichkeitstheorie verw. Gebiete*, **45,** 21–33.

Brook, D. (1964). On the distinction between the conditional and the joint probability approaches in the specification of nearest-neighbour systems. *Biometrika*, **51,** 481–483.

Buchner, M. M., Jr., and Neal, S. R. (1971). Inherent load-balancing in step-by-step switching systems. *Bell System Technical Journal*, **50,** 135–165.

Burke, P. J. (1956). The output of a queueing system. *Operat. Res.*, **4,** 699–704.

Burke, P. J. (1968). The output process of a stationary *M/M/s* queueing system. *Ann. Math. Statist.*, **39,** 1144–1152.

Burke, P. J. (1972). Output processes and tandem queues. In J. Fox (Ed.), *Proceedings of the Symposium on Computer-Communication Networks and Teletraffic*, Polytechnic Press of the Polytechnic Institute of Brooklyn, Distributed by Wiley, New York. pp. 419–428.

Chandy, K. M., Howard, J. H., and Towsley, D. F. (1977). Product form and local balance in queueing networks. *J.A.C.M.*, **24,** 250–263.

Chiang, C. L. (1968). *Introduction to Stochastic Processes in Biostatistics*, Wiley, New York.

Chung, K. L. (1967). *Markov Chains with Stationary Transition Probabilities*, 2nd ed., Springer-Verlag, Berlin.

Clifford, P., and Sudbury, A. (1973). A model for spatial conflict. *Biometrika*, **60,** 581–588.

Cohen, J. E. (1971). *Casual Groups of Monkeys and Men*, Harvard University Press, Cambridge, Mass.

Cohen, J. E. (1972). Markov population processes as models of primate social and population dynamics. *Theoret. Pop. Biol.*, **3,** 119–134.

Cohen, J. W. (1957). The generalized Engset formulae. *Philips Telecommunication Review*, **18,** 158–170.

Cohen, J. W. (1977). The multiple phase service network with generalized processor sharing. Preprint 69, Dept. of Mathematics, University of Utrecht.

Coleman, J. S., and James, J. (1961). The equilibrium size distribution of freely forming groups. *Sociometry*, **24,** 36–45.

Cooper, R. B. (1972). *Introduction to Queueing Theory*, Macmillan, New York.

Cox, D. R. (1962). *Renewal Theory*, Methuen, London.

Cox, D. R., and Miller, H. D. (1965). *The Theory of Stochastic Processes*, Chapman and Hall, London.

Cox, D. R., and Smith, W. L. (1961). *Queues*, Methuen, London.

Crow, J. F. (1972). Darwinian and non-Darwinian evolution. *Proc. Sixth Berkeley Symp. Math. Statist. Prob.*, **5,** 1–22.

Crow, J. F., and Kimura, M. (1970). *An Introduction to Population Genetics Theory*, Harper & Row, New York.

Curtois, P. J. (1977). *Decomposability: Queueing and Computer System Applications*, Academic Press, New York.

Daley, D. J. (1976). Queueing output processes. *Adv. Appl. Prob.*, **8,** 395–415.

Descloux, A. (1967). On the validity of the particular subscriber's point of view. *Proc. Fifth International Teletraffic Congress*, New York. p. 309.

Disney, R. L. (1975). Random flow in queueing networks: a review and a critique. *Trans. A.I.E.E.*, **7,** 268–288.

Ewens, W. J. (1969). *Population Genetics*, Methuen, London.

Ewens, W. J. (1972). The sampling theory of selectively neutral alleles. *Theoret. Pop. Biol.*, **3,** 87–112.

Feller, W. (1968). *An Introduction to Probability Theory and Its Applications*, 3rd ed., Wiley, New York.

Flory, P. J. (1953). *Principles of Polymer Chemistry*, Cornell University Press.

Gelenbe, E., and Muntz, R. R. (1976). Probabilistic models of computer systems—part I (exact results). *Acta Informatica*, **7,** 35–60.

Gipps, P. G. (1977). A queueing model for traffic flow. *J. Roy. Statist. Soc.*, B**39,** 276–282.

Gnedenko, B. V., and Kovalenko, I. N. (1968). *Introduction to Queueing Theory*, Israel Program for Scientific Translations, Jerusalem.

Gnedenko, B. V., and Kowalenko, I. N. (1971). *Einführung in die Bedienungstheorie*, Akademie-Verlag, Berlin.

Good, I. J. (1963). Cascade theory, and the molecular weight averages of the sol fraction. *Proc. Roy. Soc.*, A**272,** 54–59.

Goodman, L. A. (1964). Mathematical methods for the study of systems of groups. *Amer. J. Soc.*, **70,** 170–192.

Gordon, M. (1962). Good's theory of cascade processes applied to the statistics of polymer distributions. *Proc. Roy. Soc.*, A**268,** 240–259.

Gordon, W. J., and Newell, G. F. (1967). Cyclic queueing systems with restricted length queues. *Operat. Res.*, **15,** 266–278.

Grimmett, G. R. (1973). A theorem about random fields. *Bull. London Math. Soc.*, **5,** 81–84.

Gross, D., and Harris, C. M. (1974). *Fundamentals of Queueing Theory*, Wiley, New York.

Guess, H. A., and Ewens, W. J. (1972). Theoretical and simulation results relating to the neutral allele theory. *Theoret. Pop. Biol.*, **3,** 434–447.

Jackson, J. R. (1963). Jobshop-like queueing systems. *Mgmt. Sci.*, **10,** 131–142.

Jackson, R. R. P. (1954). Queueing systems with phase-type service. *Operat. Res. Quart.*, **5,** 109–120.

Kac, M. (1959). *Probability and Related Topics in Physical Sciences,* Interscience Publishers, New York.

Keilson, J. (1965). A review of transient behaviour in regular diffusion and birth–death processes, Part II. *J. Appl. Prob.*, **2,** 405–428.

Kelly, F. P. (1975a). Networks of queues with customers of different types. *J. Appl. Prob.*, **12,** 542–554.

Kelly, F. P. (1975b). Markov processes and Markov random fields. *Bull. Inst. Int. Statist.*, **46,** 397–404.

Kelly, F. P. (1976a). On stochastic population models in genetics. *J. Appl. Prob.*, **13,** 127–131.

Kelly, F. P. (1976b). Networks of queues, *Adv. Appl. Prob.*, **8,** 416–432.

Kelly, F. P. (1976c). Stochastic models with electrical analogues. *Math. Proc. Camb. Phil. Soc.*, **80,** 145–151.

Kelly, F. P. (1976d). The departure process from a queueing system. *Math. Proc. Camb. Phil. Soc.*, **80,** 283–285.

Kelly, F. P. (1977). Exact results for the Moran neutral allele model. *Adv. App. Prob.*, **9,** 197–201.

Kelly, F. P., and Ripley, B. D. (1976). A note on Strauss' model for clustering. *Biometrika*, **63,** 357–360.

Kemeny, J. G., Snell, J. L., and Knapp, A. W. (1976). *Denumerable Markov Chains,* 2nd ed., Springer-Verlag, New York.

Kendall, D. G. (1960). Unitary dilations of Markov transition operators, and the corresponding integral representations for transition-probability matrices. In U. Grenander (Ed.), *Probability and Statistics* (*The Harold Cramer Volume*), Wiley, New York.

Kendall, D. G. (1975). Some problems in mathematical genealogy. In J. Gani (Ed.), *Perspectives in Probability and Statistics: Papers in Honour of M. S. Bartlett,* Applied Probability Trust, Sheffield. Distributed by Academic Press, London, pp. 325–345.

Kent, J. T. (1978). Time reversible diffusions. *Adv. Appl. Prob.*, **10,** 819–835.

Kimura, M. (1976). Population genetics and molecular evolution. *John Hopkins Medical J.*, **138,** 253–261.

Kimura, M., and Ohta, T. (1969). The average number of generations until fixation of a mutant gene in a finite population. *Genetics*, **61,** 763–771.

Kingman, J. F. C. (1969). Markov population processes. *J. Appl. Prob.*, **6,** 1–18.

Kingman, J. F. C. (1975). Markov models for spatial variation. *The Statistician*, **24,** 167–174.

Kingman, J. F. C. (1978). Random partitions in population genetics. *Proc. Roy. Soc.*, A**361,** 1–20.

Kleinrock, L. (1964). *Communication Nets,* McGraw-Hill, New York.

Kleinrock, L. (1975). *Queueing Systems,* Vol. I, Wiley, New York.

Kleinrock, L. (1976). *Queueing Systems,* Vol. II, Wiley, New York.

Koenig, D. (1965). Verallgemeinerungen der Engsetschen formeln. *Math. Nachr.*, **28,** 145–155.

Koenig, D., and Jansen, U. (1974). Stochastic processes and properties of invariance for queueing systems with speeds and temporary interruptions. *Transactions of the Seventh Prague Conference on Information Theory, Statistical Decision Functions, and Random Processes, and of the* 1974 *European Meeting of Statisticians,* Czechoslovak Academy of Sciences. pp. 335–343.

Koenig, D., Matthes, K., and Nawrotzki, K. (1967). *Verallgemeinerungen der Erlangschen und Engsetschen Formeln* (*Eine Methode in der Bedienungstheorie*), Akademie-Verlag, Berlin.

Koenig, D., and Stoyan, D. (1976). *Methoden der Bedienungstheorie*, Akademie-Verlag, Berlin.

Koenigsberg, E. (1958). Cyclic queues. *Operat. Res. Quart.*, **9,** 22–35.

Koenigsberg, E., and Lam, R. C. (1976). Cyclic queue models of fleet operations. *Operat. Res.*, **24,** 516–529.

Kolmogorov, A. (1936). Zur Theorie der Markoffschen Ketten. *Mathematische Annalen*, **112,** 155–160.

Kovalenko, I. N. (1962). Sur la condition pour que en régime stationaire, la distribution soit indépendante des lois des durées de conversation. *Ann. des Télécommunications*, **17,** 190–191.

Lemoine, A. J. (1977). Networks of queues—a survey of equilibrium analysis. *Mgmt. Sci.*, **24,** 464–481.

Lewontin, R. C. (1974). *The Genetic Basis of Evolutionary Change*, Columbia University Press, New York.

Maher, M. J., and Cabrera, J. G. (1975). A multi-stage cyclic queueing model. *Int. J. Prod. Res.*, **13,** 255–264.

Matis, J. H., and Hartley, H. O. (1971). Stochastic compartmental analysis: model and least squares estimation from time series data. *Biometrics*, **27,** 77–102.

Mecke, J. (1975). A result on the output of stationary Erlang processes. *Math. Nachr.*, **70,** 95–98.

Melamed, B. (1979). Characterizations of Poisson traffic streams in Jackson queueing networks. *Adv. Appl. Prob.*, **11.**

Milch, P. R. (1968). A multi-dimensional linear growth birth and death process. *Ann. Math. Statist.*, **39,** 727–754.

Miller, A. J. (1961). A queueing model for road traffic flow. *J. Roy. Statist. Soc.*, B**23,** 64–90.

Moran, P. A. P. (1958). Random processes in genetics. *Proc. Camb. Phil. Soc.*, **54,** 60–72.

Moran, P. A. P. (1961). Entropy, Markov processes and Boltzmann's H theorem. *Proc. Camb. Phil. Soc.*, **57,** 833–842.

Morgan, B. J. T. (1976). Stochastic models of grouping changes. *Adv. Appl. Prob.*, **8,** 30–57.

Moussouris, J. (1974). Gibbs and Markov random systems with constraints. *J. Statist. Phys.*, **10,** 11–33.

Muntz, R. R. (1972). *Poisson Departure Processes and Queueing Networks*, IBM Research Report RC 4145, IBM Thomas J. Watson Research Centre, Yorktown Heights, New York. A shortened version of this paper appeared in *Proc. 7th Annual Conf. Information Science and Systems*, Princeton, 1973, 435–440.

Muntz, R. R. (1975). Analytic modeling of interactive systems, *Proc. I.E.E.E.*, **63,** 946–953.

Nash-Williams, C. St J. A. (1959). Random walks and electric current in networks. *Proc. Camb. Phil. Soc.*, **55,** 181–194.

Oakes, D. (1976). Random overlapping intervals: a generalization of Erlang's loss formula. *Ann. Prob.*, **4,** 940–946.

Oakes, D. (1977). Stochastic backtrackings. *Stochastic Processes Appl.*, **6,** 95–111.

Penrose, O. (1970). *Foundations of Statistical Mechanics*, Pergamon Press, Oxford.

Preston, C. J. (1974). *Gibbs States on Countable Sets*, Cambridge University Press.

Preston, C. J. (1975). Spatial birth-and-death processes. *Bull. Inst. Int. Statist.*, **46,** 371–391.

Reich, E. (1957). Waiting times when queues are in tandem. *Ann. Math. Statist.*, **28,** 768–773.

Reich, E. (1963). Note on queues in tandem. *Ann. Math. Statist.*, **34,** 338–341.
Reich, E. (1965). Departure processes. In W. L. Smith and W. Wilkinson (Eds.), *Proceedings of the Symposium on Congestion Theory*, University of North Carolina Press, Chapel Hill. pp. 439–457.
Reiser, M. (1977). Numerical methods in separable queueing networks. In M. F. Neuts (Ed.), *Algorithmic Methods in Probability*, TIMS Studies in Management Sciences, Vol. 7, North-Holland, Amsterdam. pp. 113–142.
Stockmayer, W. H. (1943). Theory of molecular size distribution and gel formation in branched chain polymers. *J. Chem. Phys.*, **11,** 45–55.
Rényi, A. (1970). *Foundations of Probability*, Holden-Day, San Francisco.
Ripley, B. D. (1977). Modelling spatial patterns. *J. Roy. Statist. Soc.*, B**39,** 172–212.
Ripley, B. D., and Kelly, F. P. (1977). Markov point processes. *J. London Math. Soc.*, **15,** 188–192.
Ross, S. M. (1970). *Applied Probability Models with Optimization Applications*, Holden-Day, San Francisco.
Ruben, H. (1964). Generalized concentration fluctuations under diffusion equilibrium. *J. Appl. Prob.*, **1,** 47–68.
Seneta, E. (1965). Quasi-stationary distributions and time reversion in genetics. *J. Roy. Statist. Soc.*, B**28,** 253–277.
Schassberger, R. (1977). Insensitivity of steady-state distributions of generalized semi-Markov processes. Part I. *Ann. Prob.*, **5,** 87–99.
Schassberger, R. (1978a). The insensitivity of stationary probabilities in networks of queues. *Adv. Appl. Prob.*, **10,** 906–912.
Schassberger, R. (1978b). *Mean Sojourn Times in Insensitive Generalised Semi-Markov Schemes*, Research Paper No. 380, University of Calgary.
Shanbhag, D. N., and Tambouratzis, D. G. (1973). Erlang's formula and some results on the departure process for a loss system. *J. Appl. Prob.*, **10,** 233–240.
Spitzer, F. (1970). Interaction of Markov processes. *Adv. Math.*, **5,** 246–290.
Spitzer, F. (1971). *Random Fields and Interacting Particle Systems*, Mathematical Association of America.
Spitzer, F. (1974). *Ecole d'Eté de Probabilités de Saint-Flour III*—1973, Springer-Verlag, Berlin.
Syski, R. (1960). *Introduction to Congestion Theory in Telephone Systems*, Oliver and Boyd, Edinburgh.
Takács, L. (1962). *Introduction to the Theory of Queues*, Oxford University Press.
Takács, L. (1969). On Erlang's formula. *Ann. Math. Stat.*, **40,** 71–78.
Taylor, J., and Jackson, R. R. P. (1954). An application of the birth and death process to the provision of spare machines. *Operat. Res. Quart.*, **5,** 95–108.
Trajstman, A. C. (1974). On a conjecture of G. A. Watterson. *Adv. Appl. Prob.*, **6,** 489–493.
Tweedie, R. L. (1978). Ed. Spatial Patterns and Processes. *Suppl. Adv. Appl. Prob.*, **10**.
Watson, G. S. (1958). On Goldberg's theory of the precipitin reaction. *J. Immunology*, **80,** 182–185.
Watterson, G. A. (1961). Markov chains with absorbing states: a genetic example. *Ann. Math. Statist.*, **32,** 716–729.
Watterson, G. A. (1974). The sampling theory of selectively neutral alleles. *Adv. Appl. Prob.*, **6,** 463–488.
Watterson, G. A. (1975). On the number of segregating sites in genetical models with recombination. *Theoret. Pop. Biol.*, **7,** 256–276.
Watterson, G. A. (1976). Reversibility and the age of an allele, I. Moran's infinitely many neutral alleles model. *Theoret. Pop. Biol.*, **10,** 239–253.

Watterson, G. A., and Guess, H. A. (1977). Is the most frequent allele the oldest? *Theoret. Pop. Biol.*, **11,** 141–160.

White, H. C. (1962). Chance models of systems of casual groups. *Sociometry*, **25,** 153–172.

White, H. C. (1970). *Chains of Opportunity*, Harvard University Press, Cambridge, Mass.

Whittle, P. (1965a). Statistical processes of aggregation and polymerization. *Proc. Camb. Phil. Soc.*, **61,** 475–495.

Whittle, P. (1965b). The equilibrium statistics of a clustering process in the uncondensed phase. *Proc. Roy. Soc.*, A**285,** 501–519.

Whittle, P. (1967). Nonlinear migration processes. *Bull. Inst. Int. Statist.*, **42,** 642–647.

Whittle, P. (1968). Equilibrium distributions for an open migration process. *J. Appl. Prob.*, **5,** 567–571.

Whittle, P. (1972). Statistics and critical points of polymerization processes. *Suppl. Adv. Appl. Prob.*, **1974,** 199–220.

Whittle, P. (1975). Reversibility and acyclicity. In J. Gani (Ed.), *Perspectives in Probability and Statistics: Papers in Honour of M. S. Bartlett,* Applied Probability Trust, Sheffield. Distributed by Academic Press, London. pp. 217–224.

Wolff, R. W., and Wrightson, C. W. (1976). An extension of Erlang's formula. *J. Appl. Prob.*, **13,** 628–632.

Symbol Index

Subject Index

Printed in the United States
by Baker & Taylor Publisher Services